T0181344

Sources and Studies
in the History of Mathematics and Physical Sciences

For other titles published in this series, go to
http://www.springer.com/series/4142

Emmy Noether (1882–1935)
(photograph 1915, Niedersächsische Staats- und Universitätsbibliothek Göttingen,
Cod. Ms. D. Hilbert 754, Nr. 73)

Yvette Kosmann-Schwarzbach

The Noether Theorems

Invariance and Conservation Laws
in the Twentieth Century

Translated by Bertram E. Schwarzbach

 Springer

Yvette Kosmann-Schwarzbach
Paris, France
yks@math.cnrs.fr

Bertram E. Schwarzbach
(Translator)

ISBN 978-0-387-87867-6   e-ISBN 978-0-387-87868-3
DOI 10.1007/978-0-387-87868-3
Springer New York Dordrecht Heidelberg London

Mathematics Subject Classification (2010): 01-02, 22-03, 22E70, 49-03, 49S05, 70-03, 70H03, 70H33, 83-03

Printed on acid-free paper

Springer is part of Springer Science+Business Media (www.springer.com)

In memory of Yseult ע"ה
who liked science as well as history

Preface

What follows thus depends upon
a combination of the methods of
the formal calculus of variations
and of Lie's theory of groups.

Emmy Noether, 1918

This book is about a fundamental text containing two theorems and their converses which established the relation between symmetries and conservation laws for variational problems. These theorems, whose importance remained obscure for decades, eventually acquired a considerable influence on the development of modern theoretical physics, and their history is related to numerous questions in physics, in mechanics and in mathematics. This text is the article "Invariante Variationsprobleme" by Emmy Noether, which was published in 1918 in the *Göttinger Nachrichten*, and of which we present an English translation in Part I of this book.

The translation of Noether's article is followed, in Part II, by a detailed analysis of its inception, as well as an account of its reception in the scientific community. As the background to Noether's research, we sketch some developments in the theory of invariants in the nineteenth century which culminated in the definition and study of differential invariants, we discuss several works in mechanics dating from the beginning of the twentieth century in which Sophus Lie's infinitesimal methods in the theory of groups began to be applied, and we show that the immediate motivation for her work was related to questions arising from Einstein's general theory of relativity of 1915. We then summarize the contents of Noether's article in modern language. In the subsequent chapters, we review the way in which Noether's contemporaries, the mathematicians Felix Klein, David Hilbert and Hermann Weyl, and the physicists Einstein and Wolfgang Pauli, acknowledged or failed to acknowledge her contribution; then we outline the quite different diffusions of her first and second theorems. Finally, we outline the genuine generalizations of Noether's results that began to appear after 1970, in the field of the calculus of variations and in the theory of integrable systems.

The present edition is based on the second edition of *Les Théorèmes de Noether. Invariance et lois de conservation au XXe siècle* (Palaiseau: Éditions de l'École Polytechnique, 2006). For this English edition, the French text has been considerably revised and augmented, with much new information and additional references.

Paris, July 2010

On the occasion of the centenary of Noether's 1918 article, this book is reprinted with corrections and very minor additions to the text and to the list of references.

Paris, July 2018

Acknowledgments

It is a pleasure to thank the many colleagues and friends who have made the writing of this book possible. When the French version first appeared in 2004, it was at the insistence of Pierre Cartier whose subsequent support was unfailing, with Alain Guichardet's generous help, and with the much appreciated contribution of Laurent Meersseman who had translated the "Invariante Variationsprobleme" into French and written a draft of the first section of a commentary.

During the preparation of the French and English texts, I received information and assistance on various questions from Alain Albouy, Sophie Bade, Henri Besson, Katherine Brading, Harvey Brown, Leo Corry, Thibault Damour, Jean Eisenstaedt, Benjamin Enriquez, Robert Gergondey, Patrick Iglésias-Zemmour, Michel Janssen, Christoph Kopper, Vitalyi Kushnirevitch, Franco Magri, Hartmann Römer, Peter Roquette, Volodya Rubtsov, David Rowe, Erhard Scholz, Steve Shnider, Reinhard Siegmund-Schultze, Jean-Marie Souriau, Jim Stasheff, Alexander Vinogradov, Salomon Wald, Stefan Waldmann and Scott Walter. To all of them—and to any colleagues whose names I may have omitted inadvertently—I am very grateful. I especially thank Peter Olver who made many useful remarks on the manuscript of the first French edition, based on his thorough knowledge of the subject, and again answered my questions when I was preparing the second edition in 2006. The three referees for the first edition of the French text made valuable recommendations and suggestions, from which I benefitted. The help I received from David Schimmel in translating several difficult German texts remains a treasured memory of this learned friend who died in a road accident in 2007.

Thanks are also due to Frédéric Zantonio at the Centre Polymédia de l'École Polytechnique who helped me with the reproductions. The editors of Springer New York, Ann Kostant, Hans Koelsch, and Elizabeth Loew, extended their support during the preparation of this book, for which I thank them cordially.

I am grateful to the institutions which have allowed me to reproduce documents in their collections: the Niedersächsische Staats- und Universitätsbibliothek, Göttingen, the Einstein archive in Jerusalem, and the archive of Bryn Mawr College, and to Princeton University Press for permission to quote from several volumes of *The Collected Papers of Albert Einstein*.

Last but not least, I thank the translator, Bertram E. Schwarzbach, for his work and for his advice on how to clarify and improve the French text.

Yvette Kosmann-Schwarzbach

July 2010 and July 2018

Contents

Part I
"Invariant Variational Problems"
by Emmy Noether
Translation of "Invariante Variationsprobleme" (1918)

Invariante Variationsprobleme.

(F. Klein zum fünfzigjährigen Doktorjubiläum.)

Von

Emmy Noether in Göttingen.

Vorgelegt von F. Klein in der Sitzung vom 26. Juli 1918[1]).

Es handelt sich um Variationsprobleme, die eine kontinuier-
liche Gruppe (im Lieschen Sinne) gestatten; die daraus sich er-
gebenden Folgerungen für die zugehörigen Differentialgleichungen
finden ihren allgemeinsten Ausdruck in den in § 1 formulierten,
in den folgenden Paragraphen bewiesenen Sätzen. Über diese aus
Variationsproblemen entspringenden Differentialgleichungen lassen
sich viel präzisere Aussagen machen als über beliebige, eine Gruppe
gestattende Differentialgleichungen, die den Gegenstand der Lieschen
Untersuchungen bilden. Das folgende beruht also auf einer Verbindung der Methoden der formalen Variationsrechnung mit denen der
Lieschen Gruppentheorie. Für spezielle Gruppen und Variationsprobleme ist diese Verbindung der Methoden nicht neu; ich er-
wähne Hamel und Herglotz für spezielle endliche, Lorentz und
seine Schüler (z. B. Fokker), Weyl und Klein für spezielle unendliche Gruppen[2]). Insbesondere sind die zweite Kleinsche Note und
die vorliegenden Ausführungen gegenseitig durch einander beein-

---

1) Die endgiltige Fassung des Manuskriptes wurde erst Ende September
eingereicht.

2) Hamel: Math. Ann. Bd. 59 und Zeitschrift f. Math. u. Phys. Bd. 50.
Herglotz: Ann. d. Phys. (4) Bd. 36, bes. § 9, S. 511. Fokker, Verslag d. Amster-
damer Akad., 27./1. 1917. Für die weitere Litteratur vergl. die zweite Note von
Klein: Göttinger Nachrichten 19. Juli 1918.

In einer eben erschienenen Arbeit von Kneser (Math. Zeitschrift Bd. 2) handelt
es sich um Aufstellung von Invarianten nach ähnlicher Methode.

First page of "Invariante Variationsprobleme" *(reproduced with permission)*
*Nachrichten von der Königlichen Gesellschaft der Wissenschaften zu Göttingen,
Mathematisch-physikalische Klasse, 1918, pp. 235–257.*

INVARIANT VARIATIONAL PROBLEMS

(For F. Klein, on the occasion of the fiftieth anniversary of his doctorate)

by **Emmy Noether** in Göttingen

Presented by F. Klein at the session of 26 July 1918[*]

We consider variational problems which are invariant[A] under a continuous group (in the sense of Lie); the consequences that are implied for the associated differential equations find their most general expression in the theorems formulated in §1, which are proven in the subsequent sections. For those differential equations that arise from variational problems, the statements that can be formulated are much more precise than for the arbitrary differential equations that are invariant under a group, which are the subject of Lie's researches. What follows thus depends upon a combination of the methods of the formal calculus of variations and of Lie's theory of groups. For certain groups and variational problems this combination is not new; I shall mention Hamel and Herglotz for certain finite groups, Lorentz and his students (for example, Fokker), Weyl and Klein for certain infinite groups.[1] In particular, Klein's second note and the following developments were mutually influential, and for this reason I take the liberty of referring to the final remarks in Klein's note.

1 Preliminary Remarks and the Formulation of the Theorems

All the functions that will be considered here will be assumed to be analytic or at least continuous and continuously differentiable a finite number of times, and single-valued within the domain that is being considered.

By the term "transformation group" one usually refers to a system of transformations such that for each transformation there exists an inverse which is an element of the system, and such that the composition of any two transformations of the system is again an element of the system. The group is called a *finite continuous* [group] $\mathfrak{G}_\rho$ when its transformations can be expressed in a general form which depends analytically on ρ *essential* parameters ε (i.e., the ρ parameters cannot be represented by ρ functions of a smaller number of parameters). In the same way, one speaks of an *infinite continuous* group $\mathfrak{G}_{\infty\rho}$ for a group whose most general transformations depend on ρ essential arbitrary functions $p(x)$ and their derivatives in a way that is

---

[*] The definitive version of the manuscript was prepared only at the end of September.

[A] *gestatten*, to permit, in the sense of admitting [an invariance group] has been translated as "being invariant under [the action of] a group" (Translator's note).

[1] Hamel, Math. Ann., vol. 59, and Zeitschrift f. Math. u. Phys., vol. 50. Herglotz, Ann. d. Phys. (4) vol. 36, in particular §9, p. 511. Fokker, Verslag d. Amsterdamer Akad., 27/1 1917. For a more complete bibliography, see Klein's second note, Göttinger Nachrichten, 19 July 1918.

In a paper by Kneser that has just appeared (Math. Zeitschrift, vol. 2), the determination of invariants is dealt with by a similar method.

Y. Kosmann-Schwarzbach, *The Noether Theorems*, Sources and Studies in the History of Mathematics and Physical Sciences, DOI 10.1007/978-0-387-87868-3_1,
© Springer Science+Business Media, LLC 2011

analytical or at least continuous and continuously differentiable a finite number of times. An intermediate case is the one in which the groups depend on an infinite number of parameters but not on arbitrary functions. Finally, one calls a group that depends not only on parameters but also on arbitrary functions a *mixed group*.[2]

Let $x_1, \ldots, x_n$ be independent variables, and let $u_1(x), \ldots, u_\mu(x)$ be functions of these variables. If one subjects the x and the u to the transformations of a group, then one should recover, among all the transformed quantities, precisely n independent variables, $y_1, \ldots, y_n$, by the assumption of invertibility of the transformations; let us call the remaining transformed variables that depend on them $v_1(y), \ldots, v_\mu(y)$. In the transformations, the derivatives of u with respect to x, that is to say $\dfrac{\partial u}{\partial x}, \dfrac{\partial^2 u}{\partial x^2}, \cdots,$ may also occur.[3] A function is said to be an *invariant* of the group if there is a relation

$$P\left(x, u, \frac{\partial u}{\partial x}, \frac{\partial^2 u}{\partial x^2}, \cdots\right) = P\left(y, v, \frac{\partial v}{\partial y}, \frac{\partial^2 v}{\partial y^2}, \cdots\right).$$

In particular, an integral I is an invariant of the group if it satisfies the relation

(1)
$$I = \int \cdots \int f\left(x, u, \frac{\partial u}{\partial x}, \frac{\partial^2 u}{\partial x^2}, \cdots\right) dx$$
$$= \int \cdots \int f\left(y, v, \frac{\partial v}{\partial y}, \frac{\partial^2 v}{\partial y^2}, \cdots\right) dy \;{}^{4}$$

integrated over an *arbitrary* real domain in x, and over the corresponding domain in y.[5]

On the other hand, I calculate for an arbitrary integral I, which is not necessarily invariant, the first variation δI, and I transform it, according to the rules of the

---

[2] Lie defines, in the "Grundlagen für die Theorie der unendlichen kontinuierlichen Transformationsgruppen" ["Basic Principles of the Theory of Infinite Continuous Transformation Groups"], Ber. d. K. Sächs. Ges. der Wissensch. 1891 (to be cited henceforth as "Grundlagen"), the infinite continuous groups as transformation groups whose elements are given by the most general solutions of a system of partial differential equations provided that these solutions do not depend exclusively on a finite number of parameters. Thus one obtains one of the above-mentioned cases distinct from that of a finite group, while, on the other hand, the limiting case of an infinite number of parameters does not necessarily satisfy a system of differential equations.

[3] I omit the indices here, and in the summations as well whenever it is possible, and I write $\dfrac{\partial^2 u}{\partial x^2}$ for $\dfrac{\partial^2 u_\alpha}{\partial x_\beta \partial x_\gamma}$, etc.

[4] I write dx, dy for $dx_1 \ldots dx_n$, $dy_1 \ldots dy_n$ for short.

[5] All the arguments $x, u, \varepsilon, p(x)$ that occur in the transformations must be assumed to be real, while the coefficients may be complex. Since the final results consist of *identities* among the x, the u, the parameters and the arbitrary functions, these identities are valid as well for the complex domain, once one assumes that all the functions that occur are analytic. In any event, a major part of the results can be proven without integration, so a restriction to the real domain is not necessary for the proof. However, the considerations at the end of §2 and at the beginning of §5 do not seem to be valid without integration.

caculus of variations, by integration by parts. Once one assumes that δu and all the derivatives that occur vanish on the boundary, but remain arbitrary elsewhere, one obtains the well-known result,

$$(2) \qquad \delta I = \int \cdots \int \delta f \, dx = \int \cdots \int \left(\sum \psi_i \left(x, u, \frac{\partial u}{\partial x}, \cdots \right) \delta u_i \right) dx,$$

where ψ represents the *Lagrangian expressions*, that is to say, the left-hand side of the Lagrangian equations of the associated variational problem $\delta I = 0$. To that integral relation there corresponds an *identity* without an integral in δu and its derivatives that one obtains by adding the boundary terms. As an integration by parts shows, these boundary terms are integrals of *divergences*, that is to say, expressions

$$\operatorname{Div} A = \frac{\partial A_1}{\partial x_1} + \cdots + \frac{\partial A_n}{\partial x_n},$$

where A is linear in δu and its derivatives. From that it follows that

$$(3) \qquad \sum \psi_i \delta u_i = \delta f + \operatorname{Div} A.$$

In particular, if f contains only the *first* derivatives of u, then, in the case of a simple integral, identity (3) is identical to Heun's "central Lagrangian equation,"

$$(4) \qquad \sum \psi_i \delta u_i = \delta f - \frac{d}{dx} \left(\sum \frac{\partial f}{\partial u_i'} \delta u_i \right), \qquad \left(u_i' = \frac{du_i}{dx} \right),$$

while for an n-fold integral, (3) becomes

$$(5) \qquad \sum \psi_i \delta u_i = \delta f - \frac{\partial}{\partial x_1} \left(\sum \frac{\partial f}{\partial \frac{\partial u_i}{\partial x_1}} \delta u_i \right) - \cdots - \frac{\partial}{\partial x_n} \left(\sum \frac{\partial f}{\partial \frac{\partial u_i}{\partial x_n}} \delta u_i \right).$$

For the simple integral and κ derivatives of the u, (3) yields

$$(6) \qquad \sum \psi_i \delta u_i = \delta f -$$

$$- \frac{d}{dx} \left\{ \sum \left(\binom{1}{1} \frac{\partial f}{\partial u_i^{(1)}} \delta u_i + \binom{2}{1} \frac{\partial f}{\partial u_i^{(2)}} \delta u_i^{(1)} + \cdots + \binom{\kappa}{1} \frac{\partial f}{\partial u_i^{(\kappa)}} \delta u_i^{(\kappa-1)} \right) \right\} +$$

$$+ \frac{d^2}{dx^2} \left\{ \sum \left(\binom{2}{2} \frac{\partial f}{\partial u_i^{(2)}} \delta u_i + \binom{3}{2} \frac{\partial f}{\partial u_i^{(3)}} \delta u_i^{(1)} + \cdots + \binom{\kappa}{2} \frac{\partial f}{\partial u_i^{(\kappa)}} \delta u_i^{(\kappa-2)} \right) \right\} +$$

$$+ \cdots + (-1)^\kappa \frac{d^\kappa}{dx^\kappa} \left\{ \sum \binom{\kappa}{\kappa} \frac{\partial f}{\partial u_i^{(\kappa)}} \delta u_i \right\},$$

and there is a corresponding identity for an n-fold integral; in particular, A contains δu and its derivatives up to order $\kappa - 1$. That the Lagrangian expressions ψ_i are actually defined by (4), (5) and (6) is a result of the fact that, by the combinations of the right-hand sides, all the higher derivatives of the δu are eliminated, while, on the other hand, relation (2), which one clearly obtains by an integration by parts, is satisfied.

In what follows we shall examine the following two theorems:

I. *If the integral I is invariant under a [group] $\mathfrak{G}_\rho$, then there are ρ linearly independent combinations among the Lagrangian expressions which become divergences— and conversely, that implies the invariance of I under a [group] $\mathfrak{G}_\rho$. The theorem remains valid in the limiting case of an infinite number of parameters.*

II. *If the integral I is invariant under a [group] $\mathfrak{G}_{\infty\rho}$ depending on arbitrary functions and their derivatives up to order σ, then there are ρ identities among the Lagrangian expressions and their derivatives up to order σ. Here as well the converse is valid.*[6]

For mixed groups, the statements of these theorems remain valid; thus one obtains identities[B] as well as divergence relations independent of them.

If we pass from these identity relations to the associated variational *problem*, that is to say, if we set $\psi = 0$,[7] then Theorem I states in the one-dimensional case— where the divergence coincides with a total differential—the existence of ρ first integrals among which, however, there may still be nonlinear identities;[8] in higher dimensions one obtains the divergence equations that, recently, have often been referred to as "conservation laws." Theorem II states that ρ Lagrangian equations are a consequence of the others.[C]

The simplest example for Theorem II—without its converse—is Weierstrass's parametric representation; here, as is well known, the integral is invariant in the case of homogeneity of the first order when one replaces the independent variable x by an arbitrary function of x which leaves u unchanged ($y = p(x)$; $v_i(y) = u_i(x)$). Thus *one* arbitrary function occurs though none of its derivatives occurs, and to this corresponds the well-known linear relation among the Lagrangian expressions themselves, $\sum \psi_i \dfrac{du_i}{dx} = 0$. Another example is offered by the physicists' "general theory of relativity"; in this case the group is the group of *all* the transformations of the $x : y_i = p_i(x)$, while the u (called $g_{\mu\nu}$ and q) are thus subjected to the transformations induced on the coefficients of a quadratic and of a linear differential form, respectively, transformations which contain the first derivatives of the arbitrary functions $p(x)$. To that there correspond the n known identities among the Lagrangian expressions and their first derivatives.[9]

---

[6] For some trivial exceptions, see §2, note 13.

[B] *Abhängigkeit*, dependence, has been translated by "identity." *Identität* has been translated by "identity" or "identity relation." Both *Relation* and *Beziehung* have been translated by "relation" and *Verbindung* by "combination" (Translator's note).

[7] More generally, one can also set $\psi_i = T_i$; see §3, note 15.

[8] See the end of §3.

[C] I.e., among the Lagrangian equations, ρ equations are consequences of the remaining ones (Translator's note).

[9] For this, see Klein's presentation.

If, in particular, one considers a group such that there is no derivative of the $u(x)$ in the transformations, and that furthermore the transformed independent quantities depend only on the x and not on the u, then (as is proven in §5) from the invariance of I, the relative invariance of $\sum \psi_i \delta u_i$[10] follows, and also that of the divergences that appear in Theorem I, once the parameters are subjected to appropriate transformations. From that it follows as well that the first integrals mentioned above are also invariant under the group. For Theorem II, the relative invariance of the left-hand sides of the identities, expressed in terms of the arbitrary functions, follows, and consequently another function whose divergence vanishes identically and which is invariant under the group—which, in the physicists' theory of relativity, establishes the link between identities and law[D] of energy.[11] Theorem II ultimately yields, in terms of group theory, the proof of a related assertion of Hilbert concerning the lack of a proper law of energy in "general relativity." As a result of these additional remarks, Theorem I includes all the known theorems in mechanics, etc., concerning first integrals, while Theorem II can be described as the maximal generalization in group theory of "general relativity."

2 Divergence Relations and Identities

Let $\mathfrak{G}$ be a continuous group—finite or infinite; one can always assume that the identity transformation corresponds to the vanishing of the parameters ε, or to the vanishing of the arbitrary functions $p(x)$,[12] respectively. The most general transformation is then of the form

$$y_i = A_i\left(x, u, \frac{\partial u}{\partial x}, \cdots\right) = x_i + \Delta x_i + \cdots$$

$$v_i(y) = B_i\left(x, u, \frac{\partial u}{\partial x}, \cdots\right) = u_i + \Delta u_i + \cdots,$$

where Δx_i, Δu_i are the terms of lowest degree in ε, or in $p(x)$ and its derivatives, respectively, and we shall assume that in fact they are *linear*. As we shall show further on, this does not restrict the generality.

---

[10] This is to say that $\sum \psi_i \delta u_i$ is invariant under the transformation up to a multiplicative factor.

[D] *Energiesatz* has been translated literally as "law of energy," in the sense of "law of conservation of energy," just as, *infra*, in §6, *eigentlich Energiesatz*, has been translated as "proper law of energy," in the sense of "proper law of conservation of energy" (Translator's note).

[11] See Klein's second note.

[12] Cf. Lie, "Grundlagen," p. 331. When dealing with arbitrary functions, it is necessary to replace the special values a^σ of the parameters by fixed functions $p^\sigma, \frac{\partial p^\sigma}{\partial x}, \cdots$; and correspondingly the values $a^\sigma + \varepsilon$ by $p^\sigma + p(x), \frac{\partial p^\sigma}{\partial x} + \frac{\partial p}{\partial x}$, etc.

Now let the integral I be invariant under $\mathfrak{G}$; then relation (1) is satisfied. In particular, I is also invariant under the infinitesimal transformations contained in $\mathfrak{G}$,

$$y_i = x_i + \Delta x_i; \qquad v_i(y) = u_i + \Delta u_i,$$

and therefore relation (1) becomes

$$
(7) \qquad 0 = \Delta I = \int \cdots \int f\left(y, v(y), \frac{\partial v}{\partial y}, \cdots\right) dy
$$
$$
- \int \cdots \int f\left(x, u(x), \frac{\partial u}{\partial x}, \cdots\right) dx,
$$

where the first integral is defined on a domain in $x + \Delta x$ corresponding to the domain in x. But this integration can be replaced by an integration on the domain in x by means of the transformation

$$
(8) \qquad \int \cdots \int f\left(y, v(y), \frac{\partial v}{\partial y}, \cdots\right) dy
$$
$$
= \int \cdots \int f\left(x, v(x), \frac{\partial v}{\partial x}, \cdots\right) dx + \int \cdots \int \operatorname{Div}(f.\,\Delta x)\, dx,
$$

which is valid for infinitesimal Δx, If, instead of the infinitesimal transformation Δu, one introduces the variation

$$
(9) \qquad \bar{\delta} u_i = v_i(x) - u_i(x) = \Delta u_i - \sum \frac{\partial u_i}{\partial x_\lambda} \Delta x_\lambda,
$$

(7) and (8) thus become

$$
(10) \qquad 0 = \int \cdots \int \{\bar{\delta} f + \operatorname{Div}(f.\,\Delta x)\} dx.
$$

The right-hand side is the classical formula for the simultaneous variation of the dependent and independent variables. Since relation (10) is satisfied by integration on an *arbitrary* domain, the integrand must vanish identically; Lie's differential equations for the invariance of I thus become the relation

$$
(11) \qquad \bar{\delta} f + \operatorname{Div}(f.\,\Delta x) = 0.
$$

If, using (3), one expresses $\bar{\delta} f$ here in terms of the Lagrangian expressions, one obtains

$$
(12) \qquad \sum \psi_i \bar{\delta} u_i = \operatorname{Div} B \qquad (B = A - f.\,\Delta x),
$$

and that relation thus represents, for each *invariant* integral I, an identity in all the arguments that occur; this is the form of Lie's differential equations for I that was sought.[13]

Let us now assume first that $\mathfrak{G}$ is a *finite continuous group* $\mathfrak{G}_\rho$; since, by assumption, Δu and Δx are linear in the parameters $\varepsilon_1, \ldots, \varepsilon_\rho$, thus, by (9), the same is true of $\bar{\delta}u$ and its derivatives; as a result, A and B are linear in ε. Therefore, if I set

$$B = B^{(1)}\varepsilon_1 + \cdots + B^{(\rho)}\varepsilon_\rho; \qquad \bar{\delta}u = \bar{\delta}u^{(1)}\varepsilon_1 + \cdots + \bar{\delta}u^{(\rho)}\varepsilon_\rho,$$

where $\bar{\delta}u^{(1)}, \ldots$ are functions of $x, u, \dfrac{\partial u}{\partial x}, \cdots$, then from (12) the desired divergence relations follow:

(13) $$\sum \psi_i \bar{\delta}u_i^{(1)} = \operatorname{Div} B^{(1)}; \ldots \qquad \sum \psi_i \bar{\delta}u_i^{(\rho)} = \operatorname{Div} B^{(\rho)}.$$

Thus ρ linearly independent combinations of the Lagrangian expressions become divergences; linear independence follows from the fact that, by (9), $\bar{\delta}u = 0, \Delta x = 0$ would imply $\Delta u = 0, \Delta x = 0$, thus also a dependence among the infinitesimal transformations. But, by assumption, such a dependence is not possible for any value of the parameters, because otherwise the group $\mathfrak{G}_\rho$, reconstructed by integration of the infinitesimal transformations, would depend on fewer than ρ essential parameters. The further possibility that $\bar{\delta}u = 0$, $\operatorname{Div}(f. \Delta x) = 0$ was also excluded. These conclusions remain valid in the limiting case of an infinite number of parameters.

Now let $\mathfrak{G}$ be *an infinite continuous group* $\mathfrak{G}_{\infty\rho}$; then once again $\bar{\delta}u$ and its derivatives, and thus also B, are linear with respect to the arbitrary functions $p(x)$ and their derivatives,[14] which yields, by introducing the values of $\bar{\delta}u$, the equation

$$\sum \psi_i \bar{\delta}u_i =$$

$$\sum_{\lambda, i} \psi_i \left\{ a_i^{(\lambda)}(x, u, \ldots) p^{(\lambda)}(x) + b_i^{(\lambda)}(x, u, \ldots) \frac{\partial p^{(\lambda)}}{\partial x} + \cdots + c_i^{(\lambda)}(x, u, \ldots) \frac{\partial^\sigma p^{(\lambda)}}{\partial x^\sigma} \right\},$$

which is independent of (12). Now, by the following identity, which is analogous to the formula for integration by parts,

$$\varphi(x, u, \ldots) \frac{\partial^\tau p(x)}{\partial x^\tau} = (-1)^\tau \cdot \frac{\partial^\tau \varphi}{\partial x^\tau} \cdot p(x) \qquad \text{mod divergences,}$$

---

[13] (12) becomes $0 = 0$ in the trivial case—which can be obtained only when $\Delta x, \Delta u$ also depend on the derivatives of u—where $\operatorname{Div}(f. \Delta x) = 0$, $\bar{\delta}u = 0$; therefore, such infinitesimal transformations are always to be removed from the group; and one takes into account only the number of remaining parameters, or of remaining arbitrary functions, in the formulation of the theorems. The question whether the remaining infinitesimal transformations still always form a group remains open.

[14] The converse will show that one does not introduce a restriction by assuming that the p are independent of the $u, \dfrac{\partial u}{\partial x}, \cdots$.

the derivatives of the p are replaced by p itself and by divergences that are linear in p and its derivatives; from that we obtain[E]

$$(14) \qquad\qquad \sum \psi_i \bar{\delta} u_i =$$

$$\sum_\lambda \left\{ (a_i^{(\lambda)} \psi_i) - \frac{\partial}{\partial x} (b_i^{(\lambda)} \psi_i) + \cdots + (-1)^\sigma \frac{\partial^\sigma}{\partial x^\sigma} (c_i^{(\lambda)} \psi_i) \right\} p^{(\lambda)} + \mathrm{Div}\,\Gamma$$

and, upon combination with (12),

$$(15)\ \sum \left\{ (a_i^{(\lambda)} \psi_i) - \frac{\partial}{\partial x} (b_i^{(\lambda)} \psi_i) + \cdots + (-1)^\sigma \frac{\partial^\sigma}{\partial x^\sigma} (c_i^{(\lambda)} \psi_i) \right\} p^{(\lambda)} = \mathrm{Div}(B - \Gamma).$$

I now form the n-fold integral of (15), extended over an arbitrary domain, and I choose the $p(x)$ so that they and all their derivatives occurring in $(B - \Gamma)$ vanish on the boundary of that domain. Since the integral of a divergence reduces to an integral on the boundary, the integral of the left-hand side of (15) vanishes as well for arbitrary $p(x)$, provided that they and sufficiently many of their derivatives vanish on the boundary; from this, by the classical rules of the calculus, the vanishing of the integrand for each $p(x)$ follows, i.e., the ρ relations[F]

$$(16)$$

$$\sum \left\{ (a_i^{(\lambda)} \psi_i) - \frac{\partial}{\partial x} (b_i^{(\lambda)} \psi_i) + \cdots + (-1)^\sigma \frac{\partial^\sigma}{\partial x^\sigma} (c_i^{(\lambda)} \psi_i) \right\} = 0 \qquad (\lambda = 1, 2, \ldots, \rho).$$

These are the identities that were sought among the Lagrangian expressions and their derivatives when I is invariant under $\mathfrak{G}_{\infty\rho}$; their linear independence can be proven as has been done above since by the converse one recovers (12), and here one can still pass from the infinitesimal transformations to the finite transformations, as we shall explain in detail in §4. Thus in a $\mathfrak{G}_{\infty\rho}$, there always exist ρ arbitrary transformations, and this is already the case for infinitesimal transformations. From (15) and (16) it again follows that $\mathrm{Div}(B - \Gamma) = 0$.

In the case of a "mixed group," if one assumes similarly that Δx and Δu are linear in the ε and the $p(x)$, one sees that, by setting the $p(x)$ and the ε successively equal to zero, divergence relations (13) as well as identities (16) are satisfied.

3 Converse in the Case of a Finite Group

To prove the converse, it is, as a first step, essentially a question of running through the previous considerations in the inverse direction. It follows from (13), after multiplication by ε and summation, that (12) is satisfied, and, by means of identity (3),

---

[E] Below, the original formula has $\sum \psi_i \delta u_i$ and (-1^σ) (Translator's note).

[F] Below, the original text reads $c_i^{(\ )}$ (Translator's note).

one obtains a relation, $\bar{\delta}f + \text{Div}(A - B) = 0$. Let us then set $\Delta x = \dfrac{1}{f} \cdot (A - B)$; one obtains (11) immediately; finally, by integration, we obtain (7), $\Delta I = 0$, which is to say, the invariance of I under the infinitesimal transformations determined by Δx and Δu, where the Δu may be calculated from Δx and $\bar{\delta}u$ by means of (9), and Δx and Δu are *linear* in the parameters. But it is well known that $\Delta I = 0$ implies the invariance of I under the finite transformations which may be obtained by integrating the system of simultaneous equations[G]

$$(17) \qquad \frac{dx_i}{dt} = \Delta x_i; \quad \frac{du_i}{dt} = \Delta u_i; \quad \begin{pmatrix} x_i = y_i \\ \\ u_i = v_i \end{pmatrix} \text{ for } t = 0 \; .$$

These finite transformations contain ρ parameters $a_1, \ldots, a_\rho$, i.e., the combinations $t\varepsilon_1, \ldots, t\varepsilon_\rho$. By the assumption of the existence of ρ and only ρ linearly independent divergence relations (13), it follows that the finite transformations always form a group if they do not contain the derivatives $\dfrac{\partial u}{\partial x}$. In the contrary case, in fact, there would be at least one infinitesimal transformation, obtained as a Lie bracket, which would not be linearly dependent on the remaining ρ; and since I remains invariant under this transformation as well, there would be more than ρ linearly independent divergence relations; otherwise this infinitesimal transformation would have the particular form $\bar{\delta}u = 0$, $\text{Div}(f \cdot \Delta x) = 0$, but, in this case, Δx or Δu would depend on derivatives, which is contrary to the assumption. The question whether this case can occur when derivatives occur in Δx or Δu is still open; it is then necessary to add all the functions Δx such that $\text{Div}(f \cdot \Delta x) = 0$ to the preceding Δx to obtain the group property, but, by convention, the supplementary parameters must not be taken into account. *Therefore the converse is proven.*

From this converse it further follows that Δx and Δu may actually be assumed to be *linear* in the parameters. In fact, if Δx and Δu were expressions of a higher degree in ε, one would simply have, because of the linear independence of the powers of ε, a greater number of corresponding relations of the type (13), from which one would deduce, by the converse, the invariance of I with respect to a group whose infinitesimal transformations depend *linearly* on the parameters. If this group must have exactly ρ parameters, then there must exist linear identities among the divergence relations originally obtained for the terms of higher degree in ε.

It still must be observed that in the case where Δx and Δu contain also derivatives of u, the finite transformations may depend on an infinity of derivatives of u; in fact, when one determines the[H] $\dfrac{d^2 x_i}{dt^2}, \dfrac{d^2 u_i}{dt^2}$, the integration of (17) leads in this case to

$$\Delta\left(\frac{\partial u}{\partial x_\kappa}\right) = \frac{\partial \Delta u}{\partial x_\kappa} - \sum_\lambda \frac{\partial u}{\partial x_\lambda}\frac{\partial \Delta x_\lambda}{\partial x_\kappa},$$ so that the number of derivatives of u increases

---

[G] Below, the original text reads $\dfrac{dx}{dt} = \Delta x_i$, then $x_i = y$ (Translator's note).

[H] The original text reads $\dfrac{{}^2 x_i}{dt^2}$ (Translator's note).

in general with every step. Here is an example:

$$f = \frac{1}{2}u'^2 ; \quad \psi = -u'' ; \quad \psi . x = \frac{d}{dx}(u - u'x) ; \quad \bar{\delta}u = x . \varepsilon ;$$

$$\Delta x = \frac{-2u}{u'^2}\varepsilon ; \quad \Delta u = \left(x - \frac{2u}{u'} \right) . \varepsilon.$$

Finally, since the Lagrangian expressions of a divergence vanish identically, the converse shows the following: if I is invariant under a $\mathfrak{G}_\rho$, then every integral which differs from I only by an integral on the boundary, which is to say, the integral of a divergence, is itself invariant under a $\mathfrak{G}_\rho$ with the same $\bar{\delta}u$, whose infinitesimal transformations will in general contain derivatives of u. Thus, in the above example, $f^* = \frac{1}{2}\left\{ u'^2 - \frac{d}{dx}\left(\frac{u^2}{x} \right) \right\}$ is invariant under the infinitesimal transformation $\Delta u = x\varepsilon, \Delta x = 0$, while in the corresponding infinitesimal transformations for f, there occur derivatives of u.

If one passes to the variational *problem*, which is to say if one lets $\psi_i = 0$,[15] then (13) yields the equations Div $B^{(1)} = 0$, ..., Div $B^{(\rho)} = 0$, which are often called "conservation laws." In the one-dimensional case, it follows that $B^{(1)} = \text{const.}, \ldots, B^{(\rho)} = \text{const.}$, and from this fact, the B contain the derivatives of order at most $(2\kappa - 1)$ of the u (by (6)) whenever Δu and Δx do not contain derivatives of an order higher than κ, the order of those derivatives that occur in f. Since, in general, the derivatives of order 2κ occur in ψ,[16] the *existence of ρ first integrals* follows. That there may be nonlinear identities among them is proven once again by the aforementioned f. To linearly independent $\Delta u = \varepsilon_1, \Delta x = \varepsilon_2$ there correspond linearly independent relations $u'' = \frac{d}{dx}u'; u''.u' = \frac{1}{2}\frac{d}{dx}(u')^2$, while there exists a nonlinear identity among the first integrals $u' = \text{const.}; u'^2 = \text{const.}$ Furthermore, we are dealing here only with the elementary case in which $\Delta u, \Delta x$ do not contain derivatives of the u.[17]

4 Converse in the Case of an Infinite Group

Let us first show that the assumption of the linearity of Δx and Δu does not constitute a restriction because, even without recourse to the converse, it is an immediate result of the fact that $\mathfrak{G}_{\infty\rho}$ depends formally on ρ and *only* ρ arbitrary functions.

---

[15] $\psi_i = 0$ or, in a slightly more general fashion, $\psi_i = T_i$, where T_i are functions recently introduced, are called in physics "field equations." In the case where $\psi_i = T_i$, the identities (13) become the equations Div $B^{(\lambda)} = \sum T_i \, \delta u_i^{(\lambda)}$, which are also called "conservation laws" in physics.

[16] Once f is nonlinear in the derivatives of order κ.

[17] Otherwise, one still obtains that $(u')^\lambda = \text{const.}$ [The original text reads u'^λ (Translator's note).] for every λ from $u''. (u')^{\lambda-1} = \frac{1}{\lambda}\frac{d}{dx}(u')^\lambda$.

One shows in fact that, in the nonlinear case, in the course of the composition of transformations whereby the terms of lower order are added, the number of arbitrary functions would increase. In fact, let

$$y = A\left(x, u, \frac{\partial u}{\partial x}, \cdots; p\right) = x + \sum a(x, u, \cdots) p^\nu + b(x, u, \cdots) p^{\nu-1} \frac{\partial p}{\partial x}$$

$$+ c p^{\nu-2} \left(\frac{\partial p}{\partial x}\right)^2 + \cdots + d \left(\frac{\partial p}{\partial x}\right)^\nu + \cdots \quad (p^\nu = (p^{(1)})^{\nu_1} \cdots (p^{(\rho)})^{\nu_\rho});$$

and corresponding to that, $v = B\left(x, u, \frac{\partial u}{\partial x}, \cdots; p\right)$; by composition with $z = A\left(y, v, \frac{\partial v}{\partial y}, \cdots; q\right)$ one obtains, for the terms of lower order,

$$z = x + \sum a(p^\nu + q^\nu) + b\left\{p^{\nu-1} \frac{\partial p}{\partial x} + q^{\nu-1} \frac{\partial q}{\partial x}\right\}$$

$$+ c\left\{p^{\nu-2} \left(\frac{\partial p}{\partial x}\right)^2 + q^{\nu-2} \left(\frac{\partial q}{\partial x}\right)^2\right\} + \cdots.$$

If any of the coefficients different from a and b is nonvanishing, one obtains in fact a term $p^{\nu-\sigma} \left(\frac{\partial p}{\partial x}\right)^\sigma + q^{\nu-\sigma} \left(\frac{\partial q}{\partial x}\right)^\sigma$ for $\sigma > 1$, which cannot be written as the differential of a *unique* function or of a power of such a function; the number of arbitrary functions would thus have increased, contrary to the hypothesis. If all the coefficients different from a and b vanish, then, according to the value of the exponents $\nu_1, \ldots, \nu_\rho$, either the second term is the differential of the first (which, for example, always occurs for a $\mathfrak{G}_{\infty 1}$) so that in fact there is linearity, or the number of arbitrary functions increases here as well. The infinitesimal transformations thus satisfy a system of linear partial differential equations because of the *linearity* of the $p(x)$; and since the group properties are satisfied, they form an "infinite group of infinitesimal transformations" according to Lie's definition (Grundlagen, §10).

The *converse* is proven by considerations similar to those of the case of finite groups. The existence of the identities (16) leads, after multiplication by $p^{(\lambda)}(x)$ and summation, and by identity (14), to $\sum \psi_i \bar{\delta} u_i = \text{Div } \Gamma$; and from there follow, as in §3, the determination of Δx and Δu and the invariance of I under infinitesimal transformations which effectively depend linearly on ρ arbitrary functions and their derivatives up to order σ. That these infinitesimal transformations, when they do not contain any derivatives $\frac{\partial u}{\partial x}, \cdots$, certainly form a group follows, as it did in §3, from the fact that otherwise, by composition, *more* than ρ arbitrary functions would occur, whereas, by assumption, there are only ρ identities (16); they form in fact an "infinite group of infinitesimal transformations." Now such a group consists (Grundlagen, Theorem VII, p. 391) of the most general infinitesimal transformations of some "infinite group $\mathfrak{G}$ of finite transformations," in the sense of Lie. Each finite transformation is generated by infinitesial transformations (Grundlagen, §7)[18] and can then be obtained by integration of the simultaneous system[I]

$$\frac{dx_i}{dt} = \Delta x_i \; ; \quad \frac{du_i}{dt} = \Delta u_i \quad \begin{pmatrix} x_i = y_i \\ \\ u_i = v_i \end{pmatrix} \quad \text{for } t = 0 \Bigg) ,$$

in which, however, it may occur that it is necessary to assume that the arbitrary $p(x)$ also depend on t. Thus $\mathfrak{G}$ *actually depends on ρ arbitrary functions*; it suffices in particular to assume that $p(x)$ is independent of t for that dependence to be analytic in the arbitrary functions $q(x) = t.p(x)$.[19] If the derivatives $\dfrac{\partial u}{\partial x}, \cdots$ are present, it may be necessary to add the infinitesimal transformation $\bar{\delta}u = 0$, $\mathrm{Div}(f . \Delta x) = 0$ in order to be able to formulate the same conclusions.

Let us add, following an example of Lie (Grundlagen, §7), a fairly general case where one can obtain an explicit formula which shows as well that the derivatives up to order σ of the arbitrary functions occur, and where the converse is thus complete. These are groups of infinitesimal transformations to which there corresponds the group of *all* the transformations of the x and those of the u "induced" by them, i.e., the transformations of the u for which Δu and therefore u only depend on those arbitrary functions that occur in Δx; there, once more, let us assume that the derivatives $\dfrac{\partial u}{\partial x}, \cdots$ do not occur in Δu. Then we have

$$\Delta x_i = p^{(i)}(x); \quad \Delta u_i = \sum_{\lambda=1}^{n} \left\{ a^{(\lambda)}(x,u)p^{(\lambda)} + b^{(\lambda)}\frac{\partial p^{(\lambda)}}{\partial x} + \cdots + c^{(\lambda)}\frac{\partial^{\sigma}p^{(\lambda)}}{\partial x^{\sigma}} \right\}.$$

Since the infinitesimal transformation $\Delta x = p(x)$ generates every transformation $x = y + g(y)$ with arbitrary $g(y)$, one can, in particular, determine $p(x)$ that depends on t in such a way that the one-parameter group will be generated by

$$(18) \qquad\qquad\qquad\qquad x_i = y_i + t.g_i(y),$$

which becomes the identity for $t = 0$, and the required form $x = y + g(y)$ for $t = 1$. In fact, from the differentiation of (18), it follows that:

$$(19) \qquad\qquad\qquad\qquad \frac{dx_i}{dt} = g_i(y) = p^{(i)}(x,t),$$

---

[18] From that it follows in particular that the group $\mathfrak{G}$ generated by the infinitesimal transformations $\Delta x, \Delta u$ of a $\mathfrak{G}_{\infty\rho}$ recovers $\mathfrak{G}_{\infty\rho}$. In fact, this $\mathfrak{G}_{\infty\rho}$ does not contain any infinitesimal transformations other than $\Delta x, \Delta u$ depending on arbitrary functions, nor can it contain any which are independent of these functions and which would depend on parameters, because it would be a case of a mixed group. Now, according to the above, the finite transformations are determined from the infinitesimal transformations.

[I] Below, the original text reads $u_i = v$ (Translator's note).

[19] The question whether this last case always occurs was raised by Lie in another formulation (Grundlagen, §7 and §13, conclusion).

where $p(x,t)$ is determined from $g(y)$ by the inversion of (18); and conversely, (18) follows from (19) because of the auxiliary condition $x_i = y_i$ for $t = 0$, by which the integral is uniquely determined. By means of (18), the x may be replaced in Δu by the "constants of integration" y and by t; moreover, the $g(y)$ and their derivatives precisely up to order σ occur in it when one expresses the $\dfrac{\partial y}{\partial x}$ in $\dfrac{\partial p}{\partial x} = \sum \dfrac{\partial g}{\partial y_\kappa} \dfrac{\partial y_\kappa}{\partial x}$ as functions of the $\dfrac{\partial x}{\partial y}$ and, in general, when one replaces $\dfrac{\partial^\sigma p}{\partial x^\sigma}$ by its expression in $\dfrac{\partial g}{\partial y}, \dots, \dfrac{\partial x}{\partial y}, \dots, \dfrac{\partial^\sigma x}{\partial y^\sigma}$. In order to determine u, one then obtains the system of equations

$$\frac{du_i}{dt} = F_i\left(g(y), \frac{\partial g}{\partial y}, \dots, \frac{\partial^\sigma g}{\partial y^\sigma}, u, t\right) \quad (u_i = v_i \text{ for } t = 0),$$

in which only t and u are variables, while the $g(y), \dots$ belong to the domain of the coefficients, so that integration yields

$$u_i = v_i + B_i\left(v, g(y), \frac{\partial g}{\partial y}, \dots, \frac{\partial^\sigma g}{\partial y^\sigma}, t\right)_{t=1},$$

which is to say, transformations that *depend on exactly σ derivatives of the arbitrary functions*. According to (18), the identity is among these transformations for $g(y) = 0$; and the group property follows from the fact that the specified procedure yields *every* transformation $x = y + g(y)$, from which the induced transformation of the u is uniquely determined, and the group $\mathfrak{G}$ is thus completely described.

A further consequence of the converse is that one imposes no restriction by assuming that the arbitrary functions depend only on x and not on $u, \dfrac{\partial u}{\partial x}, \cdots$. In fact, in this last case, in the identical reformulation (14) and also in (15), the $\dfrac{\partial p^{(\lambda)}}{\partial u}, \dfrac{\partial p^{(\lambda)}}{\partial \frac{\partial u}{\partial x}}, \cdots$ would appear in addition to the $p^{(\lambda)}$. If one then assumes successively that the $p^{(\lambda)}$ are [polynomials] of degree $0, 1, \dots$ in $u, \dfrac{\partial u}{\partial x}, \cdots$, with coefficients that are arbitrary functions of x, one simply obtains a larger number of identities (16); but, by the above converse, one returns to the preceding case by including arbitrary functions that depend exclusively on x. In the same fashion, one can prove that the simultaneous existence of identities and divergence relations which are independent of the identities corresponds to mixed groups.[20]

---

[20] As in §3, it is also a consequence of the converse that, in addition to I, any integral I^* that differs from I only by the integral of a divergence is invariant under an infinite group with the same $\bar{\delta}u$, but where Δx and Δu will in general contain derivatives of the u. Such an integral I^* was introduced by Einstein in the general theory of relativity to obtain a simpler expression for the law of energy; here I give the infinitesimal transformations that leave this I^* invariant, while retaining precisely the notation of Klein's second note. The integral $I = \int \cdots \int K\,d\omega = \int \cdots \int \mathfrak{K}\,dS$ is invariant under the group of *all* transformations of the w and those induced from them on the $g_{\mu\nu}$; to this there correspond the identities

5 Invariance of the Various Elements of the Relations

Upon restriction to the simplest case for the group $\mathfrak{G}$, the case that is usually treated, in which one does not admit any derivatives of the u in the transformations, and where the transformed independent variables depend only on x and not on u, one may conclude that the various terms in the formulas are invariant. First one deduces from known laws the invariance of $\int \cdots \int (\sum \psi_i \delta u_i)\, dx$, whence the relative invariance of $\sum \psi_i \delta u_i$,[21] where δ denotes an arbitrary variation. In fact, on the one hand,

$$\delta I = \int \cdots \int \delta f\left(x, u, \frac{\partial u}{\partial x}, \cdots\right) dx = \int \cdots \int \delta f\left(y, v, \frac{\partial v}{\partial y}, \cdots\right) dy,$$

and on the other, for a $\delta u, \delta \dfrac{\partial u}{\partial x}, \cdots$ which vanishes on the boundary and which, because of the homogeneous linear transformation of $\delta u, \delta \dfrac{\partial u}{\partial x}, \cdots$, corresponds to a $\delta v, \delta \dfrac{\partial v}{\partial y}, \cdots$ that also vanishes on the boundary:

$$\int \cdots \int \delta f\left(x, u, \frac{\partial u}{\partial x}, \cdots\right) dx = \int \cdots \int \left(\sum \psi_i(u, \ldots)\delta u_i\right) dx;$$

$$\int \cdots \int \delta f\left(y, v, \frac{\partial v}{\partial y}, \cdots\right) dy = \int \cdots \int \left(\sum \psi_i(v, \ldots)\delta v_i\right) dy,$$

$$\sum \mathfrak{K}_{\mu\nu} g_\tau^{\mu\nu} + 2\sum \frac{\partial g^{\mu\sigma} \mathfrak{K}_{\mu\tau}}{\partial w^\sigma} = 0,$$

which are equation (30) in Klein. [Above, the original text reads $\partial g^{\mu\nu}$ (Translator's note).] Now let $I^* = \int \cdots \int \mathfrak{K}^* dS$, where $\mathfrak{K}^* = \mathfrak{K} + \mathrm{Div}$ and thus $\mathfrak{K}^*_{\mu\nu} = \mathfrak{K}_{\mu\nu}$, where $\mathfrak{K}^*_{\mu\nu}$, $\mathfrak{K}_{\mu\nu}$ are the respective Lagrangian expressions. The identities derived above are also satisfied by $\mathfrak{K}^*_{\mu\nu}$; and after multiplication by p^τ and summation, one obtains, when one recognizes the differential of a product,

$$\sum \mathfrak{K}_{\mu\nu} p^{\mu\nu} + 2\,\mathrm{Div}(\sum g^{\mu\sigma} \mathfrak{K}_{\mu\tau} p^\tau) = 0\,;$$

$$\delta \mathfrak{K}^* + \mathrm{Div}\left(\sum (2g^{\mu\sigma} \mathfrak{K}_{\mu\tau} p^\tau - \frac{\partial \mathfrak{K}^*}{\partial g^{\mu\nu}_\sigma} p^{\mu\nu})\right) = 0.$$

[The original text omits the parentheses within the summation symbol (Translator's note).] Comparing the above with Lie's differential equation, $\delta \mathfrak{K}^* + \mathrm{Div}(\mathfrak{K}^* \Delta w) = 0$, one obtains

$$\Delta w^\sigma = \frac{1}{\mathfrak{K}^*} \cdot \left(\sum (2g^{\mu\sigma} \mathfrak{K}_{\mu\tau} p^\tau - \frac{\partial \mathfrak{K}^*}{\partial g^{\mu\nu}_\sigma} p^{\mu\nu})\right); \quad \Delta g^{\mu\nu} = p^{\mu\nu} + \sum g^{\mu\nu}_\sigma \Delta w^\sigma$$

[The original text omits the last parenthesis but one (Translator's note).] as infinitesimal transformations that leave I^* invariant. These infinitesimal transformations thus depend on the first and second derivatives of the $g^{\mu\nu}$, and contain the arbitrary functions p and their first derivatives.

[21] That means that $\sum \psi_i \delta u_i$ is invariant up to a factor, which is what one calls relative invariance in the algebraic theory of invariants.

and therefore, for the $\delta u, \delta \dfrac{\partial u}{\partial x}, \ldots$ that vanish on the boundary,

$$\int \cdots \int \left(\sum \psi_i(u,\ldots)\delta u_i \right) dx = \int \cdots \int \left(\sum \psi_i(v,\ldots)\delta v_i \right) dy$$

$$= \int \cdots \int \left(\sum \psi_i(v,\ldots)\delta v_i \right) \left| \frac{\partial y_i}{\partial x_\kappa} \right| dx.$$

If one expresses $y, v, \delta v$ in the third integral as functions of the $x, u, \delta u$, and if one sets this integral equal to the first integral, then one obtains the relation

$$\int \cdots \int \left(\sum \chi_i(u,\ldots)\delta u_i \right) dx = 0$$

for arbitrary δu that vanish on the boundary but are otherwise arbitrary, and from that follows, as is well known, the vanishing of the integrand for every δu; *then one obtains the relation, which is an identity in δu,*

$$\sum \psi_i(u,\ldots)\delta u_i = \left| \frac{\partial y_i}{\partial x_\kappa} \right| \left(\sum \psi_i(v,\ldots)\delta v_i \right),$$

which asserts the relative invariance of $\sum \psi_i \delta u_i$ *and, as a result, the invariance of*
$\int \cdots \int \left(\sum \psi_i \delta u_i \right) dx.$[22]

To apply this to the divergence relations and to the identities that have been obtained, it is first necessary to prove that the $\bar{\delta} u$ derived from $\Delta u, \Delta x$ actually satisfies the transformation laws for the variation δu provided that in $\bar{\delta} v$, the parameters, respectively the arbitrary functions, are determined in such a fashion that they correspond to a group similar to that of the infinitesimal transformations in y, v. Let us denote the transformation that changes x, u into y, v by $\mathfrak{T}_q$; let $\mathfrak{T}_p$ be an infinitesimal transformation in x, u; then the similar transformation in y, v is given by[J] $\mathfrak{T}_r = \mathfrak{T}_q \mathfrak{T}_p \mathfrak{T}_q^{-1}$, where the parameters, respectively the arbitrary functions r, are

---

[22] These conclusions are no longer valid when y also depends on the u, because in this case $\delta f \left(y, v, \dfrac{\partial v}{\partial y}, \cdots \right)$ also contains terms $\sum \dfrac{\partial f}{\partial y} \delta y$, and the transformation by divergences does not lead to the Lagrangian expressions, even if one neglects the derivatives of the u; in fact, in this case, the δv linear combinations of $\delta u, \delta \dfrac{\partial u}{\partial x}, \ldots$, will only lead, after a new transformation by divergences, to an identity $\int \cdots \int (\sum \chi_i(u,\ldots)\delta u_i)\, dx = 0$ [The original text reads δu (Translator's note).], so that on the right-hand side one does not obtain the Lagrangian expressions.

The question whether one can deduce from the invariance of $\int \cdots \int (\sum \psi_i \delta u_i)\, dx$ the existence of divergence relations is, according to the converse, equivalent to whether one can deduce from it the invariance of I under a group that induces the same $\bar{\delta} u$ but not necessarily the same $\Delta u, \Delta x$. In the particular case of a simple integral and f containing only first derivatives, for a finite group one may conclude from the invariance of the Lagrangian expressions that there exist first integrals (cf., for example, Engel, Gött. Nachr. 1916, p. 270).

[J] The original text reads $\mathfrak{T} = \mathfrak{T}_q \mathfrak{T}_p \mathfrak{T}_q^{-1}$ (Translator's note).

thus obtained from p and q. In formulas, this can be written

$$\mathfrak{T}_p : \xi = x + \Delta x(x,p); \qquad u^* = u + \Delta u(x,u,p);$$
$$\mathfrak{T}_q : y = A(x,q); \qquad v = B(x,u,q);$$
$$\mathfrak{T}_q \mathfrak{T}_p : \eta = A\big(x + \Delta x(x,p),q\big); \ v^* = B\big(x + \Delta x(p), u + \Delta u(p), q\big).$$

But it follows from this that $\mathfrak{T}_r = \mathfrak{T}_q \mathfrak{T}_p \mathfrak{T}_q^{-1}$, or

$$\eta = y + \Delta y(r) ; \quad v^* = v + \Delta v(r),$$

where, because of the invertibility of $\mathfrak{T}_q$, one can consider the x as functions of the y and concern oneself exclusively with the infinitesimal terms; then one obtains the identity

(20)
$$\eta = y + \Delta y(r) = y + \sum \frac{\partial A(x,q)}{\partial x} \Delta x(p) ;$$

$$v^* = v + \Delta v(r) = v + \sum \frac{\partial B(x,u,q)}{\partial x} \Delta x(p) + \sum \frac{\partial B(x,u,q)}{\partial u} \Delta u(p).$$

If one replaces here $\xi = x + \Delta x$ by $\xi - \Delta \xi$, where ξ can be expressed again as a function of x, then Δx disappears; in the same fashion, according to the first formula of (20), η becomes $y = \eta - \Delta \eta$; by this substitution, $\Delta u(p)$ is transformed into $\bar\delta u(p)$ and $\Delta v(r)$ into $\bar\delta v(r)$, and the second formula of (20) yields

$$v + \bar\delta v(y, v, \dots, r) = v + \sum \frac{\partial B(x,u,q)}{\partial u} \bar\delta u(p),$$

$$\bar\delta v(y, v, \dots, r) = \sum \frac{\partial B}{\partial u_\kappa} \bar\delta u_\kappa(x,u,p),$$

so that the transformation formulas for variations are effectively satisfied, once one assumes that $\bar\delta v$ depends only on the parameters, respectively the arbitrary functions r.[23]

Then in particular, the relative invariance of $\sum \psi_i \bar\delta u_i$ follows; and also, by (12), since the divergence relations are satisfied as well in y, v, there is relative invariance of Div B, and furthermore, by (14) and (13), the relative invariance of Div Γ and of the left-hand sides of the identities, expressed by means of the $p^{(\lambda)}$, where, in the transformed formulas, the arbitrary $p(x)$ (respectively the parameters) are always to be replaced by the r. It further follows that there is relative invariance of Div$(B - \Gamma)$, thus of a divergence of a system of functions $B - \Gamma$ that do not vanish identically and whose divergence vanishes identically.

From the relative invariance of Div B one can again, when the group is finite, draw a conclusion concerning the invariance of the first integrals. The transformation of

---

[23] It appears again that one must assume y to be independent of u, etc. for the conclusions to be valid. As an example, one may cite the $\delta g^{\mu\nu}$ and δq_ρ given by Klein, which are sufficient to describe the transformations for the variations once the p are subject to a vector transformation.

the parameters corresponding to the infinitesimal transformation will be, according to (20), linear and homogeneous, and, because of the invertibility of all the transformations, the ε will also be linear and homogeneous with respect to the transformed parameters ε^*. This invertibility is surely conserved when one sets $\psi = 0$, because no derivative of u occurs in (20). By equating the coefficients in ε^* in

$$\text{Div } B(x, u, \ldots, \varepsilon) = \frac{dy}{dx} \cdot \text{Div } B(y, v, \ldots, \varepsilon^*),$$

the $\frac{d}{dy} B^{(\lambda)}(y, v, \ldots)$ will also be homogeneous linear functions of the $\frac{d}{dx} B^{(\lambda)}(x, u, \ldots)$, so that $\frac{d}{dx} B^{(\lambda)}(x, u, \ldots) = 0$, that is, $B^{(\lambda)}(x, u, \ldots) = $ const., implies that $\frac{d}{dy} B^{(\lambda)}(y, v, \ldots) = 0$, that is $B^{(\lambda)}(y, v, \ldots) = $ const. *The ρ first integrals that correspond to a $\mathfrak{G}_\rho$ are also always invariant under this group*, which simplifies the subsequent integration. The simplest example is furnished by an f that does not depend on x, or does not depend on a u, which correspond respectively to the infinitesimal transformations $\Delta x = \varepsilon$, $\Delta u = 0$ and $\Delta x = 0$, $\Delta u = \varepsilon$. One obtains $\bar{\delta} u = -\varepsilon \frac{du}{dx}$, respectively ε, and since B is derived from f and $\bar{\delta} u$ by differentiation and by rational combinations, B is also independent of x, respectively of u, and is invariant under the corresponding groups.[24]

6 An Assertion of Hilbert

Finally, one can deduce from the above the proof of an assertion of Hilbert concerning the relationship between the lack of a proper law of energy and "general relativity" (Klein's first note, Göttinger Nachr. 1917, Response, paragraph 1) and this indeed in the more general setting of group theory.

Let an integral I be invariant under a $\mathfrak{G}_{\infty\rho}$, and let $\mathfrak{G}_\sigma$ be an arbitrary finite group obtained by specializing the arbitrary functions, thus a subgroup of $\mathfrak{G}_{\infty\rho}$. Then to the infinite group $\mathfrak{G}_{\infty\rho}$ there correspond identities (16), and to the finite group $\mathfrak{G}_\sigma$ there correspond divergence relations (13); and conversely, the existence of any divergence relations implies the invariance of I under a finite group identical to $\mathfrak{G}_\sigma$ if and only if the $\bar{\delta} u$ are linear combinations of those coming from $\mathfrak{G}_\sigma$. Thus the invariance under $\mathfrak{G}_\sigma$ cannot lead to any divergence relation other than (13). But since the

---

[24] In the case where, from the invariance of $\int (\sum \psi_i \delta u_i) \, dx$, the existence of first integrals already follows, the latter are not invariant under the entire group $\mathfrak{G}_\rho$; for example, $\int (u'' \delta u) \, dx$ is invariant under the infinitesimal transformations $\Delta x = \varepsilon_2$; $\Delta u = \varepsilon_1 + x \, \varepsilon_3$, while the first integral $u - u' x = $ const., which corresponds to $\Delta x = 0$, $\Delta u = x \, \varepsilon_3$, is not invariant under the two other infinitesimal transformations because it contains u as well as x explicitly. To this first integral there correspond infinitesimal transformations for f that contain derivatives. One thus sees that the invariance of $\int \cdots \int (\sum \psi_i \delta u_i) \, dx$ [The original text reads $\psi_i \, du_i$ (Translator's note).] is in all cases weaker than the invariance of I, which is to be remarked for the question raised in a preceding note [note 22].

existence of (16) implies the invariance of I under the infinitesimal transformations Δu, Δx of $\mathfrak{G}_{\infty\rho}$ for an *arbitrary* $p(x)$, it also implies in particular the invariance under the infinitesimal transformations of a $\mathfrak{G}_{\sigma}$ obtained by specializing, and therefore under $\mathfrak{G}_{\sigma}$. The divergence relations $\sum \psi_i \bar{\delta} u_i^{(\lambda)} = \text{Div } B^{(\lambda)}$ must therefore be consequences of the identities (16), which may also be written $\sum \psi_i a_i^{(\lambda)} = \text{Div } \chi^{(\lambda)}$, where the $\chi^{(\lambda)}$ are linear combinations of the Lagrangian expressions and their derivatives. Since the ψ occur linearly in (13) as well as in (16), the divergence relations themselves must also be *linear* combinations of the identities (16); from that fact it follows that $\text{Div } B^{(\lambda)} = \text{Div}(\sum \alpha.\chi^{(\kappa)})$; and the $B^{(\lambda)}$ themselves may be obtained linearly from the χ, that is to say, from the Lagrangian expressions and their derivatives, and from functions whose divergence vanishes identically, like the $B - \Gamma$ that appeared at the end of §2, for which $\text{Div}(B - \Gamma) = 0$, and where the divergence has, in addition, an invariance property. I call divergence relations in which the $B^{(\lambda)}$ are derived from the Lagrangian expressions and their derivatives in the manner indicated above "improper," and all the others "proper."

Conversely, if the divergence relations are linear combinations of the identities (16), and thus "improper," then the invariance under $\mathfrak{G}_{\infty\rho}$ implies the invariance under $\mathfrak{G}_{\sigma}$; and[K] $\mathfrak{G}_{\sigma}$ becomes a subgroup of $\mathfrak{G}_{\infty\rho}$. *The divergence relations corresponding to a finite group $\mathfrak{G}_{\sigma}$ are improper if and only if $\mathfrak{G}_{\sigma}$ is a subgroup of an infinite group under which I is invariant.*

Hilbert's assertion, in its original form, follows by specializing the groups. Under the term "group of translations" one designates the finite group

$$y_i = x_i + \varepsilon_i ; \quad v_i(y) = u_i(x),$$

that is,

$$\Delta x_i = \varepsilon_i, \quad \Delta u_i = 0, \quad \bar{\delta} u_i = -\sum_{\lambda} \frac{\partial u_i}{\partial x_{\lambda}} \varepsilon_{\lambda}.$$

We know that invariance under the group of translations expresses the fact that, in $I = \int \cdots \int f\left(x, u, \frac{\partial u}{\partial x}, \cdots\right) dx$, the x do not occur explicitly in f. The n associated divergence relations

$$\sum \psi_i \frac{\partial u_i}{\partial x_{\lambda}} = \text{Div } B^{(\lambda)} \quad (\lambda = 1, 2, \ldots, n),$$

are called "relations of energy" because the "conservation laws" $\text{Div } B^{(\lambda)} = 0$ associated with the variational problem correspond to the "laws of energy," and the $B^{(\lambda)}$ to the "energy components." We can then state: *Given I invariant under the group of translations, then the energy relations are improper if and only if I is invariant under an infinite group which contains the group of translations as a subgroup.*[25]

[K] The original text reads $\mathfrak{G}$ (Translator's note).

[25] The laws of energy in classical mechanics and even in the old "theory of relativity" (where $\sum dx^2$ is transformed into itself) are "proper" because no infinite group is involved.

An example of such an infinite group is provided by the group of *all* the transformations of x and the transformations induced on $u(x)$ in which only the *derivatives* of the arbitrary functions $p(x)$ occur; the group of translations arises from the specialization $p^{(i)}(x) = \varepsilon_i$; however, we cannot know whether—taking into account as well the groups obtained by modifying I by an integral on the boundary—we thus describe the most general of these groups. Induced transformations of the type just indicated can be obtained by subjecting the u to the transformations of the coefficients of a "total differential form," which is to say, of a form $\sum a\, d^\lambda x_i + \sum b\, d^{\lambda-1} x_i\, dx_\kappa + \cdots$ which contains higher-order differentials in addition to the dx; the more special induced transformations where the $p(x)$ only occur in the form of their first derivative are determined by the transformations of the coefficients of the usual differential forms $\sum c\, dx_{i_1} \ldots dx_{i_\lambda}$, and until now these were usually the only ones considered.

Another group of the type indicated above—which cannot be obtained by a transformation of coefficients because of the presence of the logarithmic term—would be the following:

$$y = x + p(x); \quad v_i = u_i + \log(1 + p'(x)) = u_i + \log \frac{dy}{dx};$$

$$\Delta x = p(x); \quad \Delta u_i = p'(x);^{26} \quad \bar{\delta} u_i = p'(x) - u_i'\, p(x).$$

Here, identities (16) become

$$\sum_i \left(\psi_i u_i' + \frac{d\psi_i}{dx} \right) = 0,$$

and the improper energy relations become[L]

$$\sum_i \left(\psi_i\, u_i' + \frac{d(\psi_i + \text{const.})}{dx} \right) = 0.$$

One of the simplest integrals that is invariant under this group is

$$I = \int \frac{e^{-2u_1}}{u_1' - u_2'} dx.$$

The most general form of I may be obtained by integrating Lie's differential equation (11):

$$\bar{\delta} f + \frac{d}{dx}(f.\, \Delta x) = 0,$$

which, by introducing the values of Δx and $\bar{\delta} u$, once one assumes that f only

---

[26] From these infinitesimal transformations, one recovers the finite transformations by the method given at the end of §4.

[L] Below, the original text reads $\sum$ (Translator's note).

depends on the first derivatives of u, may be transformed into

$$\frac{\partial f}{\partial x}p(x) + \left\{\sum \frac{\partial f}{\partial u_i} - \frac{\partial f}{\partial u_i'}u_i' + f\right\} p'(x) + \left\{\sum \frac{\partial f}{\partial u_i''}\right\} p''(x) = 0$$

(identically in $p(x)$, $p'(x)$ and $p''(x)$). This system of equations already possesses solutions for two functions $u(x)$ which actually contain derivatives, specifically,[M]

$$f = (u_1' - u_2')\Phi\left(u_1 - u_2, \frac{e^{-u_1}}{u_1' - u_2'}\right),$$

where Φ is an arbitrary function of the given arguments.

As Hilbert expresses his assertion, the lack of a proper law of energy constitutes a characteristic of the "general theory of relativity." For that assertion to be literally valid, it is necessary to understand the term "general relativity" in a wider sense than is usual, and to extend it to the aforementioned groups that depend on n arbitrary functions.[27]

[M] Below, the original text reads $(n_1' - u_2')$ (Translator's note).

[27] This confirms once more the accuracy of Klein's remark that the term "relativity" as it is used in physics should be replaced by "invariance with respect to a group." (Über die geometrischen Grundlagen der Lorentzgruppe [On the geometric foundations of the Lorentz group], Jhrber. d. d. Math. Vereinig., vol. 19, p. 287, 1910; reprinted in the phys. Zeitschrift.)

Part II
Invariance and Conservation Laws in the Twentieth Century
The Inception and Reception of the Noether Theorems

Part II
Invariance and Conservation Laws
in the Twentieth Century
The Inception and Reception of the Noether Theorems

Introduction

If the life of Emmy Noether and her work in mathematics have been the subject of numerous studies, none of them, in our opinion, has accorded her 1918 article, *Invariante Variationsprobleme* [1918c], the importance that it was to acquire because of its profundity and the diversity of the applications to which its results have lent themselves.

The article contained two theorems which were nearly forgotten within a few years of their publication, but whose influence since 1950 is hard to overestimate. The first concerned the invariance of a variational problem[1] under the action of a Lie group having a finite number of independent infinitesimal generators, the typical situation in both classical mechanics and special relativity. In this theorem, which is commonly referred to as "*the* Noether theorem," she formulated, in complete generality, the correspondence between the symmetries[2] of a variational problem and the conservation laws for the associated variational equations. It was to have important consequences for quantum mechanics, serving as a guide to the correspondence which associates conserved quantities to invariances, and it has become the basis for the theory of currents. Her second theorem dealt with the invariance of a variational problem under the action of a group involving arbitrary functions, a situation that is fundamental in general relativity and in gauge theories.

What is striking for the reader of Noether's article today is its generality. Since she not only considered groups of global symmetries but also their infinitesimal generators in the sense of Sophus Lie, she could introduce a very general concept

---

[1] The equations of classical and relativistic mechanics and physics are obtained by requiring that an action integral associated with a Lagrangian describing the system be extremal. Such equations are called variational equations or Euler–Lagrange equations. They are said to derive from a variational principle, also called an action principle. They express the vanishing of the variational derivative, also called the Euler–Lagrange derivative or Euler–Lagrange differential, of the Lagrangian. Noether calls the variational derivatives "the Lagrangian expressions."

[2] Nowadays, the expressions "the transformation T is a symmetry of the integral I" and "the integral I is invariant under the transformation T" are synonymous. Noether did not use the modern term "symmetries," but rather the expression "the integral I is an invariant of the group [of transformations] ..." or "the integral I admits the group ..." which was Sophus Lie's vocabulary.

of infinitesimal symmetry; thus she anticipated by nearly half a century the introduction of the generalized vector fields that now play an essential role in the theory of completely integrable systems, first by Harold H. Johnson in 1964 and Robert Hermann in 1965, then by several other mathematicians and physicists working independently. She also combined the methods of the nineteenth century's "formal calculus of variations," which would be reinvented and developed in the 1970s, with those of "Lie's theory of groups" which, in 1918, was still unknown to the physicists. It would be only with the intense development of quantum mechanics that occurred toward the end of the 1920s, and the publication of the books by Hermann Weyl [1928] and Eugene Wigner [1931], which were followed by Bartel van der Waerden's [1932], that physicists began to use group theory.

While her first theorem established a correspondence between invariance and conservation properties, in her second theorem, she showed that every variational problem that is invariant under a symmetry group depending on arbitrary functions possesses only "improper" conservation laws, and that such invariances give rise to identities satisfied by the variational derivatives. Noether thus emphasized an essential difference between special relativity and general relativity by showing which of her theorems was applicable to each of these theories. She concluded her article with a section in which she rendered precise and proved David Hilbert's conjecture concerning the nature of the law of conservation of energy in the general theory of relativity. In fact, she situated Hilbert's conjecture— she called it an "assertion"—in the much more general setting of invariance groups depending on arbitrary functions, and ended her text with a final footnote, altogether in the spirit of the Erlangen program,[3] in which she cited Felix Klein, who had written that the expression "relativity" should be replaced by the more general expression "invariance under a group."[4]

Noether's two theorems in pure mathematics can hardly be understood outside their historical context, i.e., the inception of the general theory of relativity in the period of great intellectual effervescence in Germany and especially in Göttingen that coincided with the war and the early years of the Weimar Republic.[5] She wrote quite explicitly in her article that questions arising from the general theory of

---

[3] In his inaugural lecture at the University of Erlangen in 1872, Felix Klein had formulated the idea that each kind of geometry was the study of the properties that remain invariant under the transformations of a particular group (Klein [1872]). See, e.g., Gray's article on "Geometry—formalisms and intuitions" in his book [1999] and Norton [1999]. This text was usually referred to, including by Klein himself, as the "Erlanger Programm." It was translated first into Italian at the suggestion of Beniamino Segre by Gino Fano, then still a student at the University of Turin, and soon there-after in the 1890s into French by Henri Eugène Padé and into English by M. W. Haskell.

[4] Klein [1910], p. 287, and *Gesammelte mathematische Abhandlungen*, vol. 1, p. 539.

[5] Noether's role has been ably described, first by Hans A. Kastrup [1987], then in David Rowe's fine article [1999]. Regarding, more generally, this period in the history of physics and mathematics, one may profitably consult the historical notes in *The Collected Papers of Albert Einstein*, the articles edited by Don Howard and John Stachel [1989], in particular Norton [1984] and Stachel [1989], those edited by Jeremy Gray [1999], Rowe's article on Einstein and Hilbert [2001], and the numerous references cited in them.

relativity were the inspiration for her research, and that her article clarifies the nature of the law of conservation of energy in that new theory. In their articles of 1917 and 1918 on the fundamental principles of physics and, in particular, on conservation laws, Klein and Hilbert said clearly that they had solicited Noether's assistance to resolve these questions, and that she proved a result which had been conjectured by Hilbert; they also acknowledged in passing that the consequences of her two theorems contributed to the elucidation of a difficult question in the general theory of relativity.[6]

The history of the reception of the two Noether theorems proved to be very curious. While the connection with general relativity persisted in the transmission of the second theorem, the motivation in physics for her research was quickly lost in the transmission of the first. Infrequent references to her results may be found in the works of Klein, Hilbert and Weyl. Her theorems figured prominently first in the article of Erich Bessel-Hagen [1921], then in the book of Roland Weitzenböck [1923], and were summarized in the treatise of Richard Courant and Hilbert [1924], but subsequently there was a nearly total silence about them until the 1950s. Since Noether's results were valid only for equations arising from a variational principle, one would think that their diffusion would have been connected to the evolution of the role of action principles in physics, which was not yet generally recognized when Noether wrote her article. While Hilbert considered variational principles to be essential for a suitable expression of the laws of physics, that view was firmly contested by Klein.[7] In a letter to Pauli in 1921, Klein reproached Hilbert with his "fanatical belief in the variational principles, the view that one can explain the reality of nature by means of purely mathematical considerations."[8] And clearly Einstein did not share Hilbert's opinion either. On 23 July 1916, he wrote to Théophile De Donder: "I must admit that, contrary to most of our colleagues, I am not at all of the opinion that every theory should be expressed in the form of a variational principle."[9] Pauli could declare several years later that "We would add, however, that it is not at all self-evident, from a physical point of view, that the physical laws should be derivable from an action principle,"[10] a remark nearly identical to a comment

---

[6] The question of energy conservation in general relativity is debated to this day. "In gravity theory, the definition of energy, momentum and angular momentum is a nontrivial problem that has a long and rich history" wrote the physicists Yuri N. Obukhov and Guillermo F. Rubilar in a 2006 paper, and a review of the question by László B. Szabados published online in 2009 listed 527 references.

[7] See Rowe [1999], pp. 201–202. See other reflections on the question of the validity of variational principles in physics in Anderson [1967], p. 344.

[8] "[...] der fanatische Glauben an die Variationsprinzipien, die Meinung, daß man durch bloßes math[ematisches] Nachdenken das Wesen der Natur erklären könne," letter from Klein to Pauli, 8 May 1921, in Pauli [1979], p. 31.

[9] "Ich muss Ihnen gestehen, dass ich im Gegensatz zu den meisten Kollegen überhaupt nicht der Ansicht bin, dass jede Theorie in die Form eines Variationsprinzips gebracht werden müsse," *Collected Papers* 8A, no. 240, p. 318; 8 (English), p. 235.

[10] "Wir möchten jedoch hinzufügen, daß es vom physikalischen Standpunkt durchaus nicht selbstverständlich ist, daß sich die Naturgesetze aus einem Variationsprinzip ableiten lassen," Pauli [1921], p. 769, and [1958], p. 201.

that Weyl had made a year earlier in a letter he wrote to Klein.[11] When, despite the reservations of these pioneers of modern physics, variational principles did emerge as the language of contemporary physical theories,[12] the importance of this part of Noether's work was finally recognized, her article began to be cited, and her name was definitively attached to the theorems and identities which she had proved in 1918.

By 1965 explicit references to "the Noether theorem" became more frequent and, by the 1970s, her name began to appear with increasing frequency in the mathematics, mechanics and physics literatures. The generalized symmetries that she introduced were rediscovered and then studied in the context of the geometric theory of integrable systems. Finally, after numerous publications of partial results that were already contained in her article while their authors claimed that they were original, genuine mathematical generalizations of her complete results began to appear.

But between the 1950s and 1980 or even later, citations of Noether's article were still almost always incomplete, which shows that many of the authors who cited it did so without having read it. One can identify a well-intentioned culprit for the incomplete transmission of Noether's article, Edward L. Hill, who, in a 1951 paper, reintroduced Noether's results to the mathematical physics community. But in order to simplify the exposition, Hill completely denatured her results, ignoring the second theorem entirely and presenting the first theorem only in its simplest particular case. For the thirty years after its publication, Hill's article remained the source through which mathematical physicists learned what they mistook to be the entire contents of Noether's article.

Even the most sympathetic articles about Noether, tributes that appeared shortly after her death, encyclopedia articles and textbooks of the history of mathematics, in general neglected this part of her mathematical contributions, restricting themselves to her fundamental role in the development of abstract algebra. Even the specialized historical and biographical articles that have been devoted to her more recently still present incomplete versions of her results on symmetries and conservation laws in the calculus of variations, and do not call attention to the influence that her article eventually exerted in contemporary mathematical physics.

In these pages, we shall examine the circumstances of the composition of Noether's article and present a short summary of its contents. We shall analyze the views of her contemporaries and those expressed after her death in 1935, describe the transmission of her ideas through a very small number of books and articles until the 1950s, and analyze the later, different histories of the reception of her two theorems. Finally, we shall outline some of the many modern developments and generalizations of her ideas that have taken place since the 1970s. We thus hope both to tell the strange story of the transmission of Noether's theorems and to assess the influence of the *Invariante Variationsprobleme* on mathematics, mechanics and physics as they have developed in the twentieth century.

---

[11] Letter of 28 December 1920, cited by Erhard Scholz [1999b], p. 272.

[12] Concerning the role of action integrals in classical physics and path integrals in quantum physics, where they are also called Feynman integrals, see, for example, DeWitt [1957], and, for a more recent treatment, Cartier and DeWitt-Morette [2006].

Chapter 1
The Inception of the Noether Theorems

Emmy Noether's two theorems on the relation between symmetries and conservation laws were a response to the mathematical problems that arose when Einstein proposed the generally covariant equations of general relativity, and when Hilbert and Klein pursued research related to the new physical theory. They served both to elucidate the problem of the conservation of the energy-momentum tensor in that new theory, and to reconcile formulations of the law of conservation of energy that had appeared, *a priori*, to be quite distinct. Her first theorem also offered a vast generalization of the conservation theorems in mechanics and in the special theory of relativity that had been known at the time. Using Lie's theory of continuous groups of transformations, she presented remarkably general results for the problem of applying the theory of differential invariants to the variational equations of physics.

1.1 From the Theory of Invariants to Special Relativity

The mid-nineteenth century was the period when the theory of invariants was created. Its origin is to be found in a problem in projective geometry, the search for a polynomial function, more generally, for a quantity defined on projective space, that would be invariant under any change of projective coordinates, which is to say, the search for a polynomial function or, more generally, a quantity that has an intrinsic geometric meaning.[1] The prototype of an algebraic invariant is the discriminant of a quadratic polynomial which remains identical to itself under a unimodular change of coordinates, i.e., one that conserves volumes. The vanishing of this discriminant corresponds to the degeneration of the associated quadratic equation. For Weyl, Arthur Cayley's "Mémoire sur les hyperdéterminants" [1846] was the founding paper for

---

[1] See Weitzenböck [1923], Study [1923], Weyl [1939], Dieudonné and Carrell [1971], Hawkins [1998], Procesi [1999], and Olver [1999]. The latter work contains 240 references to papers on invariants of which more than fifty were published before 1900. For the history of the theory of invariants, see, for example, the articles by Charles S. Fisher [1966] and Karen Hunger Parshall [1989].

Y. Kosmann-Schwarzbach, *The Noether Theorems*, Sources and Studies in the History of Mathematics and Physical Sciences, DOI 10.1007/978-0-387-87868-3_2,
© Springer Science+Business Media, LLC 2011

the theory of algebraic invariants.[2] Sylvester [1851] formulated the context in which one sought invariants. Given a "form," i.e., a homogeneous polynomial in several variables, and an "associated form," i.e., the polynomial such that its value on the variables which have undergone a linear or projective transformation is equal to that of the original polynomial evaluated on the nontransformed variables, he proposed that one seek quantities that remained unchanged under such a transformation, i.e., invariants. He introduced the concepts of covariant and contravariant substitutions to express the two ways in which the coefficients of a given form may be transformed into an associated form.[3] Thus defined, the search for the invariants of a form of given degree became a purely formal problem. Given a special class of forms, for example the binary quadratic forms, i.e., the homogeneous quadratic polynomials in two variables, the question was to find a complete list of all the algebraic invariants of a form of that class as functions of its coefficents. As early as 1858, Siegfried Aronhold and then Alfred Clebsch in 1861, Paul Gordan in 1868 and, after them, Heinrich Maschke [1900][1903] among other mathematicians, especially in Italy, developed an algorithmic method, called the symbolic method, based on the consideration of the decomposable elements in tensor products,[4] with the objective of obtaining from a known invariant for a form of a given class all that form's other invariants.

The research then turned toward the invariants of differential forms, in which case the coefficients are functions. Since the coefficients of those forms are not constant, their derivatives figure in the transformed expressions, and the invariants that are sought were called *differential invariants*. The symbolic method also worked for this type of invariant[5] but, because it appeared to be entirely calculatory, did not clarify the significance of the problem or reveal the new avenues that it in fact opened. On the one hand, it led naturally to the "absolute differential calculus," the tensor calculus and the covariant derivation of Gregorio Ricci-Curbastro and Tullio Levi-Civita[6] on manifolds,[7] because in fact, defining a tensor on a manifold amounts to

---

[2] Weyl [1939], p. 27.

[3] These two ways depend on whether one chooses to consider the coefficients of that "form" as, in modern terms, the components of a covariant or a contravariant tensor. Weitzenböck, in the preface to his book [1923], writes that a "tensor is finally nothing more than another name for what had hitherto been called a 'form'" ("Tensor ist ja schließlich nur ein anderer Name für das, was man bisher 'Form' genannt hat"), and, in chapter 5, §15, he defines "covariants" and "contravariants." Tensors had been introduced by Waldemar Voigt in 1898 in his studies on crystallography.

[4] Weyl [1939], p. 20. A modern description of the symbolic method may be found in Howe [1988], and see the indications in Hawkins [1998]. For examples of this method, see the papers and books cited above and, in particular, Weitzenböck [1923], chapter 1, §8, 10 and 13, and see Study [1923].

[5] See Wright [1908].

[6] An article by Ricci which gave a summary of his previous publications appeared in 1892. There subsequently appeared an article by Levi-Civita [1896], cited by Wright, and then the long article by Ricci and Levi-Civita in the *Mathematische Annalen* [1900]. See Weitzenböck [1923], chapter 13.

[7] Poincaré [1899], p. 6, note 1, wrote, "The word *variété* [translated here as 'manifold'] is now sufficiently well known so that I do not think it necessary to recall its definition. That is how one refers to a continuous set of points (or of systems of values): thus it is that in three-dimensional space,

defining it locally in a formulation that is invariant under a change of charts. Subsequently, this method was adapted for the determination of Poincaré's and Élie Cartan's integral invariants[8] to which the techniques of the variational calculus apply.[9] On the other hand, the search for methods that determine differential invariants led to differential equations that were invariant under the action of a group; one could therefore apply to this search the theory of continuous Lie groups of transformations[10] which permits expressing the invariance of an equation with respect to such a group, or even with respect to a local group. Lie had indeed devised a method for expressing such invariance by the vanishing of the directional derivatives, which have since been called Lie derivatives,[11] in the directions that are determined by the

---

any surface is a two-dimensional manifold and any line a one-dimensional manifold" ("Le mot *variété* est maintenant assez usité pour que je n'aie pas cru nécessaire d'en rappeler la définition. On appelle ainsi tout ensemble continu de points (ou de systèmes de valeurs) : c'est ainsi que dans l'espace à trois dimensions, une surface quelconque est une variété à deux dimensions et une ligne quelconque, une variété à une dimension"). But Élie Cartan, who had studied with Poincaré, gave a definition of an abstract manifold in 1925 and reproduced it in his *Leçons sur la géométrie des espaces de Riemann* (1928) where he wrote: "The general concept of a manifold is rather difficult to define precisely" ("La notion générale de variété est assez difficile à définir avec précision"). For the history of the concept of manifold, going back to Bernhard Riemann, see Scholz [1999a].

[8] Poincaré [1899], Cartan [1922]. (See also, *infra*, Chap. 4, p. 99, note 34.) In the introduction to his book on integral invariants, a published version of the course that he gave at the Sorbonne in Paris in 1920–1921, Cartan wrote (p. ix), "Several chapters are devoted to the rules for the calculus of the differential forms which appear under the symbols for multiple integration. [...] I propose to call them differential forms with exterior multiplication or, in short, exterior differential forms, because they obey the rules of H. Grassmann's exterior multiplication." ("Plusieurs chapitres sont consacrés aux règles de calcul des formes différentielles qui se présentent sous les signes d'intégration multiple. [...] Je propose de les appeler formes différentielles à multiplication extérieure, ou, plus brièvement, formes différentielles extérieures, parce qu'elles obéissent aux règles de la multiplication extérieure de H. Grassmann.").

[9] See Weitzenböck [1923], chapter 14.

[10] Lie and Engel [1893]. The continuous groups are now called Lie groups. Léon Autonne (1859–1916) entitled a note in the *Comptes rendus* of the Paris Academy of Sciences, "On an application of the groups of Mr. Lie" [1891]. To the best of our knowledge, the first printed mention in French of the expression "groupes de Lie" is to be found in the thesis of Arthur Tresse [1893], "On the differential invariants of continuous groups of transformations," defended 30 November of that year at the University of Paris. Tresse, who had been a student of Lie in Leipzig, wrote in his introduction, "I recall the general propositions of M. Lie regarding the groups defined by systems of partial differential equations, groups that I call *Lie groups*." ("Je rappelle les propositions générales de M. Lie, sur les groupes définis par des systèmes d'équations aux dérivées partielles, groupes que j'appelle *groupes de Lie*.") Letters from Tresse to Lie from 1892 have been conserved in which he had already proposed that term. (See Stubhaug [2000], English translation, p. 370.) In English, the expression *Lie groups* was not yet current when Tresse was writing. Wright [1908] still referred to "the theory of groups of Lie." On the emergence of the theory of Lie groups, see Hawkins [2000].

[11] Noether refers to the vanishing of a Lie derivative as "Lie's differential equation" ("die Lie'sche Differentialgleichung"). Jan Arnoldus Schouten [1954], p. 104, note 1, defines the Lie derivatives and asserts that the term was used for the first time by David van Dantzig in two notes published in the transactions of the Amsterdam Academy of Science [1932]. In fact, in his second note, van Dantzig defines the operation of Lie derivation on tensors, but he attributes the first use of the term to Władysław Ślebodziński [1931] and adds that he owes the definition that he is presenting

underlying infinitesimal group, i.e., the Lie algebra of the Lie group.[12] This point of view was used by Joseph Edmund Wright [1908] in his search for the invariants of quadratic differential forms.

The connection between the search for differential invariants and that for the quantities conserved in the time-evolution of physical systems appeared gradually and, in its complete generality, only in the article of Noether that is the subject of this study. It is a consequence of a particular case of her first theorem, that of a differential equation deriving from a variational principle with a single independent variable. Edmund T. Whittaker (1873–1956), in his treatise on dynamics [1904], attributed the discovery of the laws of conservation of both linear momentum (2nd ed., 1917, p. 59) and angular momentum (p. 60) to Newton who, on the one hand, had already observed that, in the absence of exterior forces, the center of mass of a mechanical system is either at rest or displaced in a uniform rectilinear motion and, on the other, had generalized Kepler's law of areas. Concerning the law of conservation of energy, Whittaker recognized the role of Joseph-Louis Lagrange (p. 62) who, according to Aurel Wintner,[13] knew the consequences of the Galilean invariance of the equations of motion as early as 1777. Indeed, Lagrange proposed a new method in his "General remarks on the motion of several bodies that attract one another following the law of inverse squared distances" in order to obtain those laws of conservation that were already known.[14]

Lagrange wrote in the "Avertissement" of his *Méchanique Analitique* [1788], "This treatise [...] will collect and present from a unified point of view the various principles that have been used until now to permit the solution of questions in mechanics."[15] He stated two fundamental principles of the calculus of variations,

---

to Schouten and Egbert R. van Kampen who introduced it in an article which would in fact be published in Warsaw, in the *Prace Matematyczno-Fizyczne*, in 1934 (vol. 41, pp. 1–19). We should remark that his article III, successor to the two articles which appeared in 1932, appeared in the same journal in 1934, but in English rather than in German, which demonstrates ever so clearly the impact that the Nazi seizure of power had upon the scientific community.

[12] The elements of the Lie algebra of a Lie group are the infinitesimal generators of its one-parameter subgroups. It is well known that the "infinitesimal group" introduced by Lie did not receive its modern name, "Lie algebra," until the 1930s. Nathan Jacobson writes, in the preface to his book [1962], p. v, that "it should be noted also that in these lectures [at the Institute for Advanced Study at Princeton in 1933–1934] Professor Weyl, although primarily concerned with the theory of continuous Lie groups, set the subject of Lie algebras on its own independent course by introducing for the first time the term "Lie algebra" as a substitute for "infinitesimal group," which had been used exclusively until then." According to A. John Coleman [1997], this term, which had in fact been proposed by Jacobson and adopted by Weyl after some hesitation, had first been used by Richard Brauer in his edition of the notes of Weyl's 1934–1935 course, but was not immediately adopted. Weyl wrote, "In homage to Sophus Lie such an algebra is nowadays called a *Lie algebra*" ([1939], p. 260). In the bibliography of Jacobson's book one finds the expressions *Lie Ring* and *Liescher Ring* for the articles written in German after 1935, by Walter Landherr in that year, by Ernst Witt in 1937 and by Hans Zassenhaus in 1939.

[13] Wintner [1941], p. 426.

[14] Lagrange [1777], p. 162; *Œuvres de Lagrange*, vol. 4, p. 406.

[15] "Cet ouvrage [...] réunira et présentera sous un même point de vue les différents Principes trouvés jusqu'ici pour faciliter la solution des questions de Méchanique," Lagrange [1788], p. v.

one of which is that "the known operation of integration by parts"[16] permits the elimination of the differentials of the variation.

He claimed that his analytical method for deriving "a general formula for the motion of bodies" ("une formule générale pour le mouvement des corps") yields "the general equations that contain the principles, or theorems known by the names of the conservation of kinetic energy, of the conservation of the motion of the center of mass, of the conservation of the momentum of rotational motion, or the principle of areas, and of the principle of least action."[17] He ascribed the first to Huygens (p. 183, and also p. 171), the second to Newton and to d'Alembert for a generalization (p. 185), the third to Euler, Daniel Bernoulli, and the Chevalier d'Arcy (1725– 1779) (p. 186) and the fourth, founded on the principle of Maupertuis (1698–1759), to Euler for isolated bodies [1744], then to himself for interacting bodies (p. 188). While, before Lagrange, the various conservation results had been taken to be first principles belonging to the foundations of dynamics, Lagrange viewed them as consequences of the equations of dynamics, an important shift of point of view. But there was still no explicit link with invariance properties in this first edition, although on page 415, for the equations of the top in what is now called "the Lagrange case," he derived a first integral from the consideration of what would later be called an ignorable variable.

Lagrange proposed "The simplest method to obtain the equations which determine the movement of an arbitrary system of bodies subject to arbitrary accelerating forces,"[18] and he concluded that the equation he obtained "is entirely analogous to those found by the method of variations for the determination of maxima and minima of integral formulas, and will have to be treated according to the same rules."[19] The method of maxima and minima had already figured prominently in Euler's treatise "Method for the determination of curves enjoying a property of maximum or minimum" [1744], where he wrote, in the chapter on elastic curves, that, just as the center of mass must rest at the lowest point, "the curvature of rays traveling through a transparent medium of varying density is also, *a priori*, determined by the principle that they must reach a given point in the shortest possible time."[20] Euler applied his methods to many problems and asserted that "the methods described in this book are not only of great use in analysis, but are also most helpful for the solution of

---

[16] "L'opération connue des intégrations par parties," *ibid.*, p. 56.

[17] "Les équations générales qui renferment les Principes, ou théorèmes connus sous les noms de *conservation des forces vives*, de *conservation du mouvement du centre de gravité*, de *conservation du moment de mouvement de rotation*, ou *Principe des aires*, et de *principe de la moindre quantité d'action*," *ibid.*, p. 182.

[18] "Méthode la plus simple pour parvenir aux équations qui déterminent le mouvement d'un système quelconque de corps animés par des forces accélératrices quelconques," *ibid.*, p. 216.

[19] "Cette équation est entièrement analogue à celles que l'on trouve par la méthode des variations pour la détermination des maxima et minima des formules intégrales, et il faudra la traiter suivant les mêmes règles," *ibid.*, p. 231.

[20] "Similiter curvatura radiorum per medium diaphanum variæ densitatis transeuntium, tam a priori est determinata, quam etiam ex hoc principio, quod tempore brevissimo ad datum locum pervenire debeant," Euler [1744], p. 246.

problems in physics."[21] In Euler's notation, the equation expressing the fact that an integral is stationary takes the form, first, $N - \dfrac{P' - P}{dx} = 0$, and then $N - \dfrac{dP}{dx} = 0$,[22] an equation to be found later also in his "Elements of the calculus of variations" [1766], and which would be generalized by Lagrange. In particular, in his letter of 1756 to Euler, Lagrange considered the variation of double integrals for the first time.

It is only in the second edition, *Mécanique Analytique* [1811], that Lagrange observed a correlation between symmetries and the principles of conservation of certain quantities, in particular energy. In the first section of the second part of his treatise, concerning dynamics, he presented a detailed history of the diverse "principles or theorems" discovered by Galileo, Huygens, Newton, Daniel Bernoulli, Euler, d'Alembert and several other physicists. Concerning the conservation of the angular momenta he wrote, "Regarding the movement of several bodies about a fixed center, the sum of the products of the mass of each of those bodies by the velocity of its motion about that center, and by its distance from that center [...] is constant so long as there is no other action nor any exterior obstacle."[23] In article 7 of the fourth section, Lagrange introduced (p. 288) the kinetic energy, $T = \dfrac{1}{2}m\left(\left(\dfrac{dx}{dt}\right)^2 + \left(\dfrac{dy}{dt}\right)^2 + \left(\dfrac{dz}{dt}\right)^2\right)$, and, in the case where the force derives from a potential,[24] which he denoted by V, he wrote, for the "Lagrangian" $T - V$, the "Euler–Lagrange equations" (article 10, p. 290) using the method of the calculus of variations which he had introduced as early as 1760 to serve as the fundamental method of dynamics.[25] Then he asserted (article 14),

An integration which can always be performed when the forces are functions of distances and the functions T, V, L, M, etc.[26] do not contain the finite variable t is the one that yields the principle of the conservation of kinetic energy.[27]

---

[21] "Methodi in hoc libro traditæ, non solum maximum esse usum in ipsa analysi, sed etiam eam ad resolutionem prolematum physicorum amplissimum subsidium afferre," *ibid.*, p. 245.

[22] Setting $N = \dfrac{\partial L}{\partial y}$ and $P = \dfrac{\partial L}{\partial y'}$, this equation takes the usual form of the case of a one-dimensional variational problem. The literature on the history of the calculus of variations is vast. See Goldstine [1980], Kreyszig [1994], and René Taton on the relations of Euler and Lagrange [1983].

[23] "Dans le mouvement de plusieurs corps autour d'un centre fixe, la somme des produits de la masse de chaque corps par sa vitesse de circulation autour du centre, et par sa distance au même centre [...] se conserve la même tant qu'il n'y a aucune action ni aucun obstacle extérieur." We cite the first volume of the 1965 edition, p. 227. We thank Professors Jean-Marie Souriau and Patrick Iglesias-Zemmour for calling our attention to several passages in Lagrange's work. We also benefited from unpublished research on Lagrange by Alain Albouy. For this aspect of Lagrange's work, see Vizgin [1972]. See also Marsden and Ratiu [1999], pp. 231–234.

[24] Modern notational practice has retained Lagrange's V for the potential which is the opposite of the force function.

[25] Lagrange [1760].

[26] $L = 0, M = 0$, etc. represent the constraint equations.

[27] "Une intégration qui a toujours lieu lorsque les forces sont des fonctions de distances [i.e., ne dépendent pas des vitesses], et que les fonctions T, V, L, M, etc., ne contiennent point la variable finie t, est celle qui donne le principe de la conservation des forces vives," p. 295.

By means of the formula for integration by parts, he then demonstrated this "principle," that is, the theorem asserting that the total energy of a system, $T + V$, remains constant.[28] Concerning the other first integrals that were already known, Lagrange was less precise, merely saying, "The other integrals will depend on the nature of the differential equations of each problem, and one cannot provide a general method for finding them."[29]

Some thirty years later, Carl Gustav Jacobi (1804–1851), in his "Lectures on Dynamics," a course given at the Universiy of Königsberg in 1842–1843,[30] dealt with the relation between the Euclidean invariance of the Lagrangian in mechanics under the action of translations and rotations, and the laws of the conservation of linear and angular momenta. The third, fourth, fifth and sixth lectures of this course deal respectively, with the principle of the conservation of the motion of the center of mass, of the kinetic energy, of areas, and with the principle of least action.[31]

In 1897, Ignaz R. Schütz, then a member of the Institute for Theoretical Physics at Göttingen,[32] studied the principle of the conservation of energy and showed that it was largely independent of the principle of the equality of action and reaction asserted by Newton, and then derived the law of conservation of energy from the equations of motion, first for an isolated massive point particle, and then for a system of particles.

It was by using the theory of Lie groups and, in particular, the concept of infinitesimal transformation, that Georg Hamel[33] proposed establishing relations between mechanics and several domains of mathematics including, in particular, the calculus of variations. He published his habilitation thesis [1904a] and then an article, "On virtual displacements in mechanics" [1904b], where he studied the equivalence of various forms of the equations of mechanics and how they would change under virtual displacements. To that end he used the Lie brackets of infinitesimal symmetries (p. 425), which he called "the Jacobi symbols" ("die Jacobischen Symbole"), as well as the structure constants of the Lie group with which he was dealing

---

[28] Concerning the meanings attributed to the conservation of energy before Hermann von Helmholtz (1821–1894) [1887] and especially Lagrange's concept of energy, consult Elkana [1974].

[29] "Les autres intégrales dépendront de la nature des équations différentielles de chaque problème ; et l'on ne saurait donner de règle générale pour les trouver," p. 297.

[30] Jacobi [1866]. This series of lectures was published posthumously by Clebsch.

[31] "Das Princip der Erhaltung der Bewegung des Schwerpunkts, der lebendigen Kraft, der Flächenräume, der kleinsten Wirkung (des kleinsten Kraftaufwandes)." The French "forces vives" and the German "lebendige Kraft" are translations of the Latin term "vis viva," introduced by Leibniz. The kinetic energy is one-half of the *vis viva*. In his book on the stability of motion [1877], Routh called the kinetic energy the "semi vis viva.'

[32] For Schütz, see Scott Walter's thesis, "Hermann Minkowski et la mathématisation de la relativité restreinte, 1905–1915," Nancy, 1996, or Rowe [2009]. Schütz, who was assistant to Ludwig Boltzmann (1844–1906) in Munich from 1891 to 1894, died in 1926. Schütz's article [1897] would be cited by Hermann Minkowski in his lecture in Cologne in 1908, translated in Lorentz *et al.* [1923].

[33] Hamel (1877–1954) was a student of Hilbert who defended his thesis in 1901. He was the author of several important treatises on mechanics. On p. 4, note 4, of [1904a], and on p. 417 of [1904b] he wrote of *der Lieschen Gruppentheorie*.

(p. 428). Ultimately, he asserted the equivalence of two forms of the equations of mechanics in the case of n virtual displacements corresponding to the infinitesimal transformations of an n-parameter group.

Next, it was Gustav Herglotz (1881–1953) who studied various questions in the mechanics of solid bodies from the point of view of the special theory of relativity [1911]. He considered the ten-parameter invariance group[34] which acts on the four-dimensional space-time, now called Minkowski space-time. In his section 9 (pp. 511–513), using a method of the calculus of variations that would be used by Noether seven years later, he derived ten first integrals associated to the ten infinitesimal transformations of the Poincaré group. This section would be cited by Noether [1918c] and by Klein [1927].

In 1916 there appeared in the *Göttinger Nachrichten* a letter that Friedrich Engel[35] had addressed to Klein in which he remarked that, working from Herglotz's result and letting the speed of light tend to infinity, one could recover the ten well-known integrals of nonrelativistic mechanics. He then proposed to obtain the same result directly, without passing to the limit, by means of Lie's theory.[36] Using the Hamiltonian formalism and the invariance of the Hamiltonian under the action of the ten infinitesimal transformations of the ten-parameter group, called the Galilean group, he obtained the ten first integrals of the n-body problem, and in particular he recovered Schütz's 1897 result on the conservation of the total energy of the system. In a second letter [1917], Engel showed how to use the conserved quantities to integrate the equations of mechanics by the method of Lie, but he did not use a variational method in either paper.

Finally, on 15 August 1918, while Noether was completing the definitive version of her manuscript of the *Invariante Variationsprobleme*, Alfred Kneser[37] submitted an article to the *Mathematische Zeitschrift*, "Least action and Galilean

---

[34] This 10-dimensional Lie group, which is the semi-direct product of the 6-dimensional Lorentz group and the 4-parameter group of translations, was called by Herglotz "the 10-term group of 'motions'" ("die zehn gliedrige Gruppe der ‚Bewegungen'"). It is now called the Poincaré group, a term used for the first time by Wigner in 1939 (see Mehra [1974], p. 70). Wigner wrote ([1967], p. 18), "I like to call the group formed by these invariables [*sic*] the Poincaré group," and referred to Poincaré's publications of the years 1905 and 1906. According to Klein (in a letter to Pauli in 1921, see Appendix III, pp. 159–160), it was Poincaré who had perceived that the transformations introduced by Lorentz form a group, and, according to Wigner ([1967], p. 5) and Pais ([1982], p. 21), it was also Poincaré who gave their name to the Lorentz transformations. In the physics literature, the 10-dimensional Poincaré group is also often called the inhomogeneous Lorentz group or, sometimes, the Lorentz group.

[35] Engel (1861–1941) had written his *Habilitationsschrift* with Lie in Leipzig in 1885 and continued publishing on group theory. He is mostly known for his work with Lie on what became the three-volume treatise, Lie and Engel [1893]. See Hawkins [2000], pp. 77–78.

[36] Engel [1916]. See Mehra [1976], pp. 70–71, note 130.

[37] Kneser (1862–1930) was a well-known specialist in integral equations and the calculus of variations. (See Thiele [1997].) The author of a monograph on the calculus of variations [1900] that was re-issued in 1925, he was also the author of the first part of the chapter on this topic in Klein's *Encyklopädie der mathematischen Wissenschaften*. He was a *Privatdocent* in Breslau (present-day Wrocław), then a professor from 1886 to 1889 at the University of Dorpat (now Tartu in Estonia), and later in Berlin, returning eventually to Breslau.

relativity" [1918] in which he developed Schütz's results [1897] using Lie's infinitesimal transformations and, as Noether would do, emphasized the relevance of Klein's Erlangen program, but did not treat questions of invariance. Slightly earlier, he had published another article [1917] where he applied the theory of Lie and Georg Scheffers to a study of variational equations and of the Hamilton–Jacobi equation, but in neither article did he touch on the problem of conserved quantities.

One can thus say that scattered results in classical and relativistic mechanics tying together properties of invariance and conserved quantities had already appeared in the publications of Noether's predecessors, without any of them having discovered the general correspondence principle. Noether supplied this general theory and consequently, after 1918, the earlier results became special cases of her first theorem. In the conclusion of his 1916 letter, Engel emphasized that the detour effected by considering the inhomogeneous Lorentz group was necessary to justify the existence of "the integral of kinetic energy and of the second integrals of the center of mass" ("das Integral des lebendigen Kraft und die zweiten Schwerpunktsintegrale") which had previously appeared "to have fallen from the heavens" ("wie vom Himmel gefallen"). Noether showed on the contrary that considering a symmetry group that was well adapted to the problem would render the known conservation laws natural, and also provided a general method for calculating conservation laws from invariances of a variational integral, and conversely, for calculating the symmetries of a variational problem from its known conservation laws.

1.2 The General Theory of Relativity and the Problem of the Conservation of Energy

The history of the discovery of general relativity has been amply studied, most recently in volumes of the series *Einstein Studies* and in the articles cited above. We shall therefore summarize only the elements of that history that are essential for an understanding of the role that Noether played in it.

In an article on the consequences of the principle of relativity, Einstein [1907] already observed that the laws of physics did not permit a distinction between a reference frame in a constant gravitational field and a uniformly accelerated reference frame, and he considered the question of the extension of the principle of relativity to this more general situation. After 1912 he sought an expression for the laws of gravitation that would be invariant under a group of transformations that would be larger than the group composed of the Lorentz transformations and translations, and would be invariant with respect to an arbitrary change of coordinates.

After several attemps in this direction and exchanges with Max Abraham and Gunnar Nordström in particular,[38] Einstein undertook, with the help of his friend, the mathematician Marcel Grossmann, a study of Ricci's and Levi-Civita's absolute

---

[38] For a detailed account of this period in Einstein's career, see Mehra [1974], Pais [1982], pp. 208–216 and 229, Rowe [1999] [2001], and the numerous references which are cited there.

differential calculus in order to supply a mathematical framework for the extension of the principle of relativity that he was seeking. He tried to formulate the laws of gravity in the form of generally covariant, second-order differential equations, which is to say, independently of the coordinate system that may be chosen, in terms of a nonconstant metric, $g_{\mu\nu}$, that would describe the gravitational potential. Einstein then temporarily abandoned the requirement that the equations of gravitation be generally covariant, because such a formulation did not yield a conservation law for energy.[39] At first he restricted his search to linear transformations; then he introduced the idea of systems of adapted coordinates which turned out to be systems of coordinates related by unimodular transformations, that is, transformations whose Jacobian equals 1 and which thus conserve volumes. This first version of general relativity is known as the *Entwurf* and is only a sketch of the eventual theory.

By restricting his search to these changes of coordinates, Einstein succeeded in November 1915 in establishing equations for gravitation. Still better, he recognized that, with a slight modification, these equations would be tensorial, thus generally covariant. On 4, 11 and 18 November 1915, he presented his conclusions before the Royal Prussian Academy of Sciences in Berlin [1915].

These new equations, however, created a grave problem because the law of the conservation of energy implied that when one adopted a suitably adapted system of coordinates, which was permitted by general covariance, the energy-momentum tensor vanished at every point in space.[40] This further implied that the scalar energy was constant. But that hypothesis was satisfied only in the case of a homogeneous gravitational field. In fact, these equations still lacked the trace term that Einstein introduced in his article of 25 November 1915, in which the equations of gravitation would find their definitive form. However, there still remained one point that was not satisfactory. The law of conservation of energy did not seem to be a direct consequence of the equations describing gravitation, nor did it seem to have mathematical justification.

---

[39] The question of the conservation of energy was among the most important of Einstein's concerns throughout his career, as can be seen from his *Annalen der Physik* articles of 1906 and 1907 dealing with the inertia of energy, as well as from his letters to Michele Besso (Einstein and Besso [1972]). In particular, see the letters written during a visit to Ahrenshoop in Pomerania, 29 July 1918, no. 45, p. 129 (*Collected Papers* 8B, no. 591, pp. 835–837; 8 (English), pp. 613–614), where he writes that the total energy of a system is "an *integral* invariant without a corresponding *differential* invariant" ("*Integral*invariante, der keine *Differential*invariante entspricht"), and 20 August 1918, no. 46, p. 132 (*Collected Papers* 8B, no. 604, pp. 858–861; 8 (English), pp. 629–630), where he argues against one of Weyl's hypotheses and returns to the question of energy by insisting on the necessity of introducing "the tension tensor for the static gravitational field" ("das Spannungstensor für das statische Gravitationsfeld"). In his introduction to this correspondence, Pierre Speziali also mentions (p. li) the letters of 28 July 1925 (from Geneva), no. 76, p. 209, and 2 August 1925 (from Berne), no. 77, p. 211, but in fact Einstein wrote about the energy tensor as early as the end of 1913 or the beginning of 1914 (letter from Zurich, no. 9, p. 51; *Collected Papers* 5, no. 499, pp. 588–589; 5 (English), pp. 373–374). In the following letter, no. 10, p. 53 (*Collected Papers* 5, no. 514, pp. 603–604; 5 (English), pp. 381–382), written from Zurich in early March 1914, Einstein evokes the "law of conservation" (*Erhaltungssatz*) together with the gravitation equations to obtain conditions on the coefficients of the metric.

[40] See Earman and Glymour [1978].

Other papers, some by such highly reputed physicists as Paul Ehrenfest (1880–1933) and Hendrik A. Lorentz (1853–1928), contributed to a clarification of the question of the conservation of energy,[41] and there were many publications related to this problem. In 1916 an article by Ehrenfest [1916] appeared in the *Proceedings* of the Royal Academy of Sciences in Amsterdam in which he calculated the invariants of a variational problem. In the same volume Lorentz proposed a Lagrangian and established the equations of general relativity from the corresponding variational principle, then derived from them the law of the conservation of momentum and energy, but this was still in the framework of the preliminary version of the general theory of relativity. In 1917, Lorentz's student Adriaan Daniel Fokker published an invariant method for obtaining those results [1917], and discussed the consequences of the variational principle. This was shortly before Weyl [1917] succeeded in deriving the theorem of energy-momentum from Hamilton's principle. Still in 1917, Nordström, citing Einstein [1916a], Herglotz [1916] and the publications of Lorentz in 1915, calculated the "tension-energy tensor of matter" ("spannings-energietensor der materie"). From March to June 1916, Lorentz delivered a series of lectures in Leiden on Einstein's theory, and published in that year and in early 1917 a series of four articles in which he presented an invariant geometric theory of general relativity [1916].[42]

Noether was to refer to "Lorentz and his students (for example Fokker)," and would explicitly cite the latter's 1917 article. She was also to refer to Weyl, but without a precise reference to any of his publications. Her second theorem unifies certain of the results of the research of her predecessors, and it is she who brought to the fore the existence of identities satisfied by the Euler–Lagrange equations which appear with an infinite-dimensional symmetry group such as the group of all transformations of the manifold of general relativity.

1.3 The Publications of Hilbert and Klein on General Relativity

Since mid 1915 Hilbert had been working intensely to understand Einstein's papers and had sought to deduce the laws of physics in a generally covariant form from a limited number of axioms by combining Gustav Mie's (1868–1957) theory of electromagnetism (1912) with Einstein's theory of gravitation.[43] Hilbert was interested in these problems because he had already proved several fundamental

---

[41] For the historical context, see Pais [1987], Sauer [1999], Cattani and De Maria [1993], and see Trautman [1962] for a very clear exposition of the difficulties posed by the problem of the conservation of energy in general relativity. (See, *infra*, Chap. 6, p. 126.) For subsequent developments, see, for example, Havas [1990].

[42] Histories of these discoveries, together with analyses of the articles in which they were announced, have been published by Michel Janssen [1992] and Anne J. Kox [1992].

[43] On the events of 1915–1918 and the scientific relations among Einstein, Hilbert, Klein, and Noether, see Rowe [1999], who provides a detailed analysis based on archival documents. See also Einstein's correspondence in *Collected Papers* 8A.

theorems concerning invariants, and because relativity entered into the outstanding questions about geometry that had perplexed both Klein and himself. Already in the 1872 Erlangen program, Klein had defined a geometry as the data of a manifold and a group of transformations of that manifold, in modern terminology, a group of diffeomorphisms, thus identifying the study of a geometry with the search for the invariants of that group. Hilbert clearly saw a connection between, on the one hand, the theory of invariants and geometry, and, on the other, the problem of extending the special theory of relativity.

In late June–early July 1915, Einstein came to Göttingen at Hilbert's invitation[44] to deliver a series of lectures on the general theory of relativity—which was still the preliminary version which he would discard in November of that year. He was so enthusiastic about Hilbert and his reception of his theory that he wrote to his friend Heinrich Zangger upon his return on 7 July, "I was one week in Göttingen and learnt to know and like him. I delivered there six two-hour lectures on the now well clarified theory of gravitation, and I had the pleasure of completely convincing the mathematicians there [in Göttingen],"[45] and to Arnold Sommerfeld on 15 July, "In Göttingen I had the great pleasure to see that everything was understood to the last detail. I am most delighted with Hilbert."[46]

Hilbert and Einstein conducted an intense correspondence during the months of October and November 1915 in which they developed their closely related theories. Hilbert's approach was different from Einstein's because he used a variational principle to obtain the field equations.[47] In fact, in his article dated 20 November 1915, Hilbert introduced two axioms and a generally invariant function from which he deduced ten gravitational equations and four electromagnetic equations, all of which were covariant with respect to any change of coordinates. As the study of the proofs of Hilbert's article [1915] has demonstrated, his gravitational equations were Einstein's equations [1915] of which he had been apprised in a letter that he received when his article was still in proof, so that Einstein indeed had priority in

---

[44] From Einstein's correspondence, we know the dates of his stay in Göttingen, from 26 or 27 June to 5 July, since he wrote to Hilbert on 24 June that he would call on him on Monday morning (28 June 1915) (*Collected Papers* 8A, no. 91, p. 142; 8 (English), p. 107). A letter of 6 July mentions that he had returned from Göttingen the previous night (letter to Wander and Geertruida de Haas, *ibid.*, no. 92, pp. 142–143; 8 (English), p. 108).

[45] "Ich war eine Woche in Göttingen wo ich ihn kennen und lieben lernte. Ich hielt dort sechs zweistündige Vorträge über die nun schon sehr geklärte Gravitationstheorie und erlebte die Freude, die dortigen Mathematiker vollständig zu überzeugen," *Collected Papers* 8A, no. 94, pp. 144–145; 8 (English), pp. 109–110.

[46] "In Göttingen hatte ich die grosse Freude, alles bis ins Einzelne verstanden zu sehen. Von Hilbert bin ich ganz begeistert," *Collected Papers* 8A, no. 96, p. 147; 8 (English), p. 111. This quotation is also translated by Mehra [1974], p. 25. See Pais [1982], p. 259.

[47] One of Einstein's manuscripts, entitled "Appendix: Formulation of a theory on the basis of a variational principle," has now been published in the *Collected Papers* 6, no. 31, pp. 340–346. According to the editors, this text, which was written before 20 March 1916, may have been intended to serve as the last section of, or as an appendix to, his long article [1916a]. It was at the end of 1916 that Einstein published an article on a variational formulation of general relativity [1916b]. For more details regarding variational formulations of Einstein's equations, see Kishenassamy [1993]. Cf. also Noether [1918c], pp. 249–250, note 1 (pp. 15–16, note 20, in the above translation).

the discovery of the equations that bear his name.[48] By applying a theorem that he stated without proof, Hilbert obtained a conservation law for the energy-momentum tensor which, at first glance, was different from Einstein's. That theorem would be proved three years later by Noether.[49]

During the years 1917 and 1918, Klein and Einstein corresponded frequently, and the problem of the conservation of energy was the subject of numerous comments and requests for explanations that preceded and followed the publication of their several articles.[50] Klein and Hilbert also exchanged letters in 1918 about the conservation of energy and related topics. It is known that Klein discussed the problem of the conservation of energy with Noether and also with Carl Runge[51] in the spring of 1918, and that, together with Runge, he undertook a systematic study of the bibliography of the subject. Klein wrote to Hilbert on 5 March 1918,[52] informing him that he had spoken before the Royal Scientific Society in Göttingen (*Königliche Gesellschaft der Wissenschaften zu Göttingen*) on 25 February, advocating that one consider only the energy tensor of matter, and not that of gravitation, in the energetic balance of a field, that Runge had further developed his, i.e., Klein's, idea of the energetic balance of the gravitational field, that Runge would develop it "very well" ("sehr schön") the coming Friday (8 March 1918) in a lecture before the Scientific Society, and inviting him to attend, "Do come on Friday evening to the Scientific Society."[53] He added that Runge had put this theorem in a regular form by a suitable choice of coordinates for each particular case. Hilbert replied on 7 March, sending proofs of his "first note," in which he "worked out directly Runge's ideas."[54] In fact, on 8 March, Runge delivered a lecture before the Scientific Society, "On the Theorem of the Conservation of Energy in Gravitational Theory" ("Über den Satz von der Erhaltung der Energie in der Gravitationstheorie"). As early as 12 March, Noether wrote to Klein criticizing Runge's ideas.[55] In his letter to Einstein of

---

[48] See Corry, Renn and Stachel [1997]. The problems of priority of discovery and of the relations between Hilbert and Einstein were first studied by Mehra [1974], then by Earman and Glymour [1978], Pais [1982], pp. 257ff. and 274–275, and Vizgin [1994], chapter 2. Also see Rowe [1999], pp. 199–205, and [2001], and the historical notes in *The Collected Papers of Albert Einstein*, 8A.

[49] See Section 6 of Noether's article (pp. 19–22 in the above translation) and, *infra*, Chap. 2, pp. 63–64.

[50] Klein [1918a, b and c] and Einstein [1916b] and [1918].

[51] Runge (1856–1927) published on very diverse subjects during his career. He had been a full professor of applied mathematics at Göttingen since 1904.

[52] The two letters concerning the law of conservation of energy that Klein wrote to Hilbert on 5 February and 5 March 1918, and Hilbert's brief answer, have been published in Hilbert and Klein [1985], nos. 126, 128 and 129, pp. 140–144.

[53] "Kommen Sie doch ja am Freitag Abend noch in die Gesellschaft der Wissenschaften," *ibid.*, no. 128, p. 142.

[54] "[...] meiner ersten Mitteilung, in der ich gerade die Ideen von Runge auch ausgeführt hatte," *ibid.*, no. 129, p. 144. This "first note" does not correspond to any of Hilbert's published articles and may have been a draft. Hilbert refused to attend Runge's lecture to protest the presence of Edward Schröder (a professor of German philology at Göttingen) on the board of directors of the Scientific Society (*ibid.*).

[55] See Appendix II, pp. 153–157, and in particular note 4.

20 March 1918,[56] Klein mentioned the results that Runge had obtained, and informed him that his paper and Runge's were nearly ready for publication but, on 24 March, Einstein argued against Runge's ideas,[57] and that convinced both Klein and Runge not to publish the papers that they had been preparing "until [they] had arrived at a better perspective on the entire literature" dealing with the subject.[58] In early June, Klein proposed to speak about the article that Einstein was about to publish, "The theorem of the conservation of energy in general relativity,"[59] and, in his letter of 9 June 1918, Einstein wrote to Klein, "I am very pleased that you will talk about my article on energy. I shall now give you a complete proof of the tensorial character (for linear transformations) of J_σ."[60] But Klein finally abandoned the projected lecture because he was not convinced of the validity of Einstein's argument.

In early 1918 Klein published an article [1918a][61] in the form of an exchange of letters with Hilbert in which he simplified the argument that Hilbert had published in his article on "The foundations of physics" [1915], and offered a discreet criticism of that article. The uncertainties about the relationship beween the theories of Einstein and Hilbert, in particular regarding the associated laws of conservation, were finally dissipated by Klein in his later articles of 1918, "On the differential laws for the conservation of momentum and energy in Einstein's theory of gravitation" [1918b] (19 July) and "On the integral form of conservation laws and the theory of the spatially closed universe" [1918c] (6 December), where he elucidated, with Noether's theorems playing an essential role,[62] the derivation of Einstein's and Hilbert's laws of conservation and the vectorial nature of the quantities that Hilbert had defined. But the principal difficulty, that of explaining the difference in the nature of the conservation laws in classical mechanics and special relativity on the one hand, and in general relativity on the other hand, had, in fact, already been resolved by Noether in March 1918, and explained in the article [1918c] that was presented by Klein at the Scientific Society in July and submitted for publication in September of that year. Now, Einstein was evidently not aware of this immediately because he could still write to Klein, on 13 March, "The relations here [in general relativity] are exactly

---

[56] Einstein, *Collected Papers* 8A, no. 487, pp. 685–690; 8 (English), pp. 503–507.

[57] *Collected Papers* 8B, no. 492, pp. 697–699; 8 (English), pp. 512–514.

[58] "[...] wenn wir die volle Uebersicht über die jetzt vorliegende Literatur haben," letter of 18 May 1918 from Klein to Einstein (*Collected Papers* 8B, no. 540, pp. 761–762; 8 (English), p. 559). Preliminary versions of the paper that Klein was preparing but which he chose not to publish as well as notes of his discussions with Runge have been conserved in the Göttingen archives (see Einstein, *Collected Papers* 7, p. 76, note 5, on Einstein's article [1918]). Runge never returned to this question after learning of Einstein's criticisms of his project.

[59] Einstein [1918].

[60] "Es freut mich sehr, dass Sie über meine Energie-Arbeit vortragen werden. Ich teile Ihnen nun den Beweis für den Tensorcharakter (bez. linearer Transformationen) von J_σ vollständig mit," *Collected Papers* 8B, no. 561, p. 791; 8 (English), p. 581.

[61] Even though it appeared in the volume dated 1917, Klein's article was actually submitted to the journal 25 January 1918.

[62] See, *infra*, for a more detailed analysis of the chronology of Noether's discoveries, pp. 46–48, and for Hilbert's and Klein's acknowledgments of Noether's contribution, in Chap. 3, pp. 66–71.

analogous to those of nonrelativistic theories."[63] We shall see (p. 47) that around that date, Noether, who was then visiting in Erlangen, had already written to Klein about this very point, which explains why, on 20 March, Klein could assert to Einstein that what he had claimed was far from being true. Einstein replied once again that one could consider the fact that the integrals $\int (\mathfrak{T}_\sigma^4 + t_\sigma^4)\, dV$ are constant with respect to time "as being entirely analogous and equivalent to the conservation law for the energy-momentum in the classical mechanics of continua."[64] What Noether had contributed to the question of Hilbert's energy vector was essential, as Klein would write to Einstein on 10 November 1918,[65] the only time in his correspondence with Einstein that he mentions Noether and the importance of her contribution.

Much later, in 1924, it was Schouten and Dirk Struik who observed that, in the special case of the Lagrangian of general relativity, the identities obtained by Noether's second theorem were also consequences of the Bianchi identities, which were well known in Riemannian geometry. They express the vanishing of the covariant differential of the curvature of the Levi-Civita connection associated to a metric.[66]

The difficult problem of the conservation of energy in general relativity began to be understood much later when the gravitation theory was put into Hamiltonian form by Richard Arnowitt, Stanley Deser and Charles W. Misner in 1962. There remained the problem of proving the positivity of the energy, which was eventually achieved by Edward Witten in 1981.[67]

1.4 Emmy Noether at Göttingen

Emmy Amalie Noether (1882–1935), "of Bavarian nationality and Israelite confession,"[68] was the daughter of the mathematician Max Noether (1844–1921). She

---

[63] "Es liegen hier genau analoge Verhältnisse vor wie bei den nicht-relativistischen Theorien," *Collected Papers* 8B, no. 480, p. 673; 8 (English), p. 494.

[64] "[...] welche dem Impuls-Energie-Satz der klassichen Mechanik der Kontinua als durchaus gleichartig und gleichwertig an die Seite gestellt werden kann," letter of 24 March 1918 cited in note 57. The symbol t_σ^4 denotes the time-components of the quantities t_σ^ν which Einstein had introduced in his article [1916b] and about which he complained to Hilbert in a letter of 12 April 1918, "everybody rejects my t_σ^ν as though they were not kosher"! ("Meine t_σ^ν werden als unkoscher von allen abgelehnt," *Collected Papers* 8A, no. 503, p. 715; 8 (English), p. 525). For the "pseudo-tensor" t_σ^ν, see, *infra*, Chap. 6, p. 127.

[65] Einstein, *Collected Papers* 8B, no. 650, p. 942; 8 (English), p. 692. See, *infra*, Chap. 3, p. 70.

[66] These identities were named after the Italian geometer Luigi Bianchi (1856–1928). See Levi-Civita [1925], p. 182, where the history of the Bianchi identities is sketched, and see Pais [1982], chapter 15c, pp. 274–278. In fact, in a 1917 article where he introduced the idea of parallel displacement, Levi-Civita had already applied the contracted Bianchi identities to the theory of gravitation, and had corresponded with Einstein on the subject. See Cattani and De Maria [1993] and Rowe [2002].

[67] See Faddeev [1982], Choquet-Bruhat [1984].

[68] "[...] bayerische-Staatsangehörigkeit und israelitische-Konfession," as she described herself in the beginning of a manuscript *curriculum vitæ* written around 1917 and reproduced on the first

wrote her doctoral thesis at Erlangen in 1907 under Paul Gordan (1837–1912), one of the most distinguished specialists in the theory of invariants. In Erlangen she also came under the influence of Ernst Fischer[69] (1875–1954). Her thesis, "On the Construction of the System of Forms of a Ternary Biquadratic Form," which dealt with the search for the invariants of a ternary biquadratic form, i.e., of a homogeneous polynomial of degree 4 in 3 variables, was published in "Crelle's Journal" [1908], while an extract had appeared a year earlier [1907]. In her next article, "On the theory of invariants of forms of n variables" [1911], which had been announced the year before its publication (Noether [1910]), she extended the arguments of her thesis to the case of forms in n variables. Then she studied the fields of rational functions in "Fields and systems of rational functions" [1915] which she had announced in the *Jahresbericht der Deutschen Mathematiker-Vereinigung* (Noether [1913]). She had joined the German Mathematical Society (*Deutsche Mathematiker-Vereinigung*, or DMV) in 1909.

In 1916, in volume 77 of the *Mathematische Annalen*, Noether published a series of three articles [1916a, b, c] and then a fourth [1916d] on algebraic invariants. Regarding her articles on the invariants of finite groups[70] and on the search for bases of invariants that furnish expansions with integral or rational coefficients [1916a, b], Weyl wrote in 1935,

> The proof of finiteness is given by her for the invariants of a finite group (without using Hilbert's general basis theorem for ideals), for invariants with restriction to integral coefficients, and finally she attacks the same question along with the question of a minimum basis consisting of independent elements, for the fields of rational functions.[71]

In his book on the classical groups that appeared four years later, Weyl gave a summary of the proof contained in Noether [1916a],

> An elementary proof [of the first main theorem] for finite groups not depending on Hilbert's general theorem on polynomial ideals was given by E. Noether.[72]

And still later, in his analysis of Hilbert's work, he cited that article in a footnote once more.[73]

---

page of her *Gesammelte Abhandlungen / Collected Papers*. A mention of religious affiliation was normally part of one's national identity in Germany in that period. For the biography of Noether, see Dick [1970] [1981], Kimberling [1981] and Srinivasan and Sally [1983]. A relatively complete electronic bibliography of materials relating to her life and works with links to other pertinent sites may be found at the web-site of the association « femmes et mathématiques ».

[69] On this point, see Weyl [1935a].

[70] For a modern version of the results of Noether [1916a] and an account of developments in the theory of invariants of finite groups, see Smith [2000], and for an extension to the case of prime charateristic of her results on a bound for the degrees of the generators of the ring of polynomial invariants for finite groups, see Fogarty [2001]. Noether herself had considered the case of prime charateristic in 1926.

[71] Weyl [1935a], p. 206, *Gesammelte Abhandlungen*, vol. 3, p. 430. This eulogy by Weyl, in English, was quoted in its entirety by Dick [1970], pp. 53–72, and [1981], pp. 112–152. See, *infra*, Chap. 3, pp. 77–78.

[72] Weyl [1939], p. 275. The reference to Noether's article is on p. 314, note 19 of chapter 8.

[73] Weyl [1944], p. 621, and *Gesammelte Abhandlungen*, vol. 4, p. 139, note 2. In Reid [1970], pp. 245–283, Weyl's 1944 text is abridged but includes the note referring to Noether (p. 249).

Noether's next publication [1918a] dealt with equations that admit a prescribed Galois group, a study that extends her 1915 article.[74]

In 1915 Klein and Hilbert invited Noether to Göttingen to help them in the development of the implications of general relativity theory, and she arrived in the spring. Research in the Göttingen archives[75] has shown that Noether took an active part in Klein's seminar. The seminars in Berlin in those years, despite Einstein's presence there, were much less oriented toward mathematical physics and, in particular, toward the mathematics of relativity theory.[76] The list of themes treated in Klein's seminar has been published in the Supplement to volume 3 of his *Gesammelte mathematische Abhandlungen*, p. 11. We extract from it the following titles:
– Summer 1916, Theory of invariants of linear transformations,
– Winter 1916/17, Theory of special relativity on an invariant basis,
– Summer 1917, Theory of invariants of general point transformations,
– Summer 1918–Winter 1918/19 until Christmas, General theory of relativity on an invariant basis, [...]
– Winter 1920/21 until Christmas, Variational principles of classical mechanics and of general relativity.
Shortly after her arrival in Göttingen, Noether began work on the problem of the invariants of differential equations, and, in 1918, she published two articles on the subject, "Invariants of arbitrary differential expressions" [1918b] and "Invariant variational problems," *Invariante Variationsprobleme* [1918c], the article that will be studied here.[77] In it she takes up the work initiated by Hamel [1904a, b] and Herglotz [1911]. At the request of Hilbert, some time before May 1916, she had begun to study the various problems that resulted from the formulation of the general theory of relativity, and it is clear from a letter from Hilbert to Einstein of 27 May 1916 that she had already written some notes on the subject, notes that have not yet been identified and may not have been conserved. Hilbert wrote, "My law [of conservation] of energy is probably linked to yours; I have already given Miss Noether this question to study." In the next sentence he explained why the vectors a^l and b^l that had been considered by Einstein could not vanish in the limiting case in which the coefficients of the metric are constant, and he added that, to avoid a long explanation, he had appended to his letter "the enclosed note of Miss Noether."[78] On

---

[74] According to the algebraist Paul Dubreil [1986], this problem had been posed by Richard Dedekind (1831–1916). Works utilizing Noether's results and conjectures on this question have been analyzed by Richard G. Swan in the section "Galois Theory" of the chapter "Noether's Mathematics" in Brewer and Smith [1981], pp. 115–124.

[75] Rowe [1999].

[76] "Berlin is no match for Göttingen, in what concerns the liveliness of scientific interest, at all events in this area" ("Berlin kann sich, was Lebhaftigkeit des wissenschaftlichen Interesses anbelangt, wenigstens auf diesem Gebiete mit Göttingen nicht messen," letter of 7 July 1915 from Einstein to Heinrich Zangger, cited in note 45 above).

[77] The 1918 volume of the *Göttinger Nachrichten* which contains these two articles is available at the site http://www.emani.org (SUB Göttingen).

[78] "Mein Energiesatz wird wohl mit dem Ihrigen zusammenhängen: ich habe Frl. Nöther diese Frage schon übergeben. [...] Ich lege der Kürze [Wegen] den beiliegenden Zettel von Frl. Nöther bei," *Collected Papers* 8A, no. 222, pp. 290–292; 8 (English), pp. 215–216.

30 May 1916 Einstein answered him in a brief letter, "[...] I now understand everything in your article except the energy theorem." He then derived from the equation that Hilbert had proposed an apparently absurd consequence "which would deprive the theorem of its sense," and then asked, "How can this be clarified?" and continued, "Of course it would be sufficient if you asked Miss Noether to clarify this for me."[79] This exchange shows that Noether's expertise in this area of the discussions concerning general relativity was conceded by both Hilbert and Einstein as early as her first year in Göttingen.

Although her work on the energy vector that had been introduced by Hilbert began in 1916,[80] it was in the winter and spring of 1918 that Noether discovered the profound reason for the difficulties that had arisen in the interpretation of the conservation laws in general relativity. These considerations would be clearly stated in the *Invariante Variationsprobleme*, which contains two theorems on the relationship between the group of transformations that leave invariant the action integral of a Lagrangian system and the conservation laws, the one in the case of an invariance group with a finite number of parameters, the situation in classical mechanics and special relativity, and the other in the case of an invariance group of the same type as the group that figures in general relativity, a generally covariant theory, which is to say, one whose field equations are invariant under any change of coordinates. Thus what distinguishes the two cases is the presence in the second case of an invariance group depending on arbitrary functions.

It is on the *verso* of a postcard that Noether addressed to Klein from Erlangen, 15 February 1918,[81] that she sketched her second theorem. The formula in her line 8,

$$\delta f - \frac{\partial}{\partial x_1} \sum_i \frac{\partial f}{\partial \frac{\partial z_i}{\partial x_1}} \delta z_i - \cdots \frac{\partial}{\partial x_n} \sum_i \frac{\partial f}{\partial \frac{\partial z_i}{\partial x_n}} \delta z_i = - \sum_i \psi_i(z)\, \delta z_i,$$

is, except for some slight changes in notation and the sign convention adopted for the quantities ψ_i, identical to formula (5) of her article. In that article, equation (5) is preceded by equation (3) which contains the definition of the components ψ_i of the Euler–Lagrange derivative—she calls them the "Lagrangian expressions" ("die Lagrangeschen Ausdrücke")—of the Lagrangian f and which introduces the divergence term Div A. Then formula (5) provides the explicit expression of the quantity A in the case of n independent variables and a first-order Lagrangian.

Further on, the long equation that occupies two lines corresponds to the case of invariance under each of the translations of an n-dimensional space which, in the case of special relativity, is the 4-dimensional Minkowski space. Noether therefore considers, for every $\kappa = 1, 2, \ldots, n$, the variation $\delta z_i = \frac{\partial z_i}{\partial x_\kappa}$ which implies that, if f

[79] "In Ihrer Arbeit ist mir nun verständlich ausser dem Energiesatz. [...] was dem Satze seinen Sinn rauben würde. Wie klärt sich dies? Es genügt ja, wenn Sie Frl. Nöther beauftragen, mich aufzuklären," *Collected Papers* 8A, no. 223, pp. 293–294; 8 (English), pp. 216–217.

[80] See also the passages in Klein and Hilbert that are cited *infra*, Chap. 3, p. 65, as well as Mehra [1974], p. 70, note 129a, and Rowe [1999], p. 213.

[81] See a reproduction, as well as the transcription and a translation in Appendix I, pp. 149–151.

does not depend explicitly on x_κ, the variation of f is the total derivative of f with respect to x_κ. She then obtains "the n identities" which appear on two lines in the middle of the page,

$$\frac{\partial}{\partial x_1}\left(\sum_i \frac{\partial f}{\partial \frac{\partial z_i}{\partial x_1}}\frac{\partial z_i}{\partial x_\kappa}\right)+\cdots+\frac{\partial}{\partial x_\kappa}\left(\sum_i \frac{\partial f}{\partial \frac{\partial z_i}{\partial x_\kappa}}\frac{\partial z_i}{\partial x_\kappa}-f\right)+\cdots\frac{\partial}{\partial x_n}\left(\sum_i \frac{\partial f}{\partial \frac{\partial z_i}{\partial x_n}}\frac{\partial z_i}{\partial x_\kappa}\right)$$

$$=\sum_i \psi_i(z)\frac{\partial z_i}{\partial x_\kappa};\quad (\kappa=1,2\ldots n).$$

She has thus determined the n components of the n conserved currents, i.e., the n vector fields whose divergence vanishes when the Euler–Lagrange equations are satisfied, associated with the n spatial directions.[82] In the case of special relativity, these $n^2=16$ components are those of the energy-momentum tensor.

But, in a generally covariant theory on an n-dimensional space which, in the case of general relativity, is a curved space-time with $n=4$ dimensions, the space-time admits all the changes of coordinates where x'_κ is an arbitrary function of the x_λ's, which corresponds to an infinitesimal symmetry where $\dfrac{\partial}{\partial x_\kappa}$ is multiplied by an arbitrary function of the x_λ's. From that Noether deduces that, in the generally covariant case, the identities

$$\sum_i \psi_i(z)\frac{\partial z_i}{\partial x_\kappa}=0;\quad (\kappa=1,2\ldots n)$$

are satisfied by the Lagrangian expressions, which shows that "the ρ equations, $\psi_i=0$, are equivalent to $\rho-n$ [equations]." Those identities appear four lines above the end of the text of Noether's postcard. As we shall emphasize (Chap. 2, p. 61), these identities are special cases of the general formula (16) that she would prove in the second section of her article. She writes here that she "hopes to be able to prove the general case, where the scalars z_α are replaced by the tensors $g_{\mu\nu}$, in an analogous manner," which shows that a solution of the problem posed by the general theory of relativity was already in view.

A month later, in her letter to Klein of 12 March 1918,[83] Noether formulated the fundamental idea that the lack of a theorem concerning energy in general

---

[82] If one introduces the shorthand notation z^i_λ for $\dfrac{\partial z_i}{\partial x_\lambda}$, the components of the conserved current associated with the infinitesimal symmetry $\dfrac{\partial}{\partial x_\kappa}$ are thus $N_1^{(\kappa)},\ldots,N_n^{(\kappa)}$, where

$$N_\lambda^{(\kappa)}=-\sum_{i=1}^{i=n}\frac{\partial f}{\partial z^i_\lambda}z^i_\kappa+f\delta_{\kappa\lambda},$$

with $\delta_{\kappa\lambda}=1$ if $\kappa=\lambda$ and 0 otherwise. In [1918c], Noether would introduce the variation $\bar{\delta}z_i$ which is, in this case, $-\dfrac{\partial z_i}{\partial x_\kappa}$, because the vector field $\dfrac{\partial}{\partial x_\kappa}$ has components $\delta_{\kappa\lambda}$, $\lambda=1,2,\ldots,n$.

[83] See a reproduction of this letter, its transcription and a translation in Appendix II, pp. 153–157.

relativity is due to the fact that the invariance groups that were considered were in fact subgroups of an infinite group, and therefore led to identities that are satisfied by the Lagrangian expressions, "by my additional research, I have now established that the [conservation] law for energy is not valid in the case of invariance *under any extended group generated by the transformation induced by the z's*."[84] Here z designates the set of dependent variables, and the last words of the emphasized sentence should be understood as "invariance under the transformations of the z's induced by *all* the transformations of the independent variables." A comparison of this sentence with the wording in Noether's section 6 shows that this is a preliminary formulation of an essential consequence of what would become her second theorem.

On 23 July 1918 Noether delivered a paper before the Göttingen Mathematical Society (*Mathematische Gesellschaft zu Göttingen*)[85] entitled, like the eventual article, *Invariante Variationsprobleme*, and whose summary begins, "In connection with research related to Hilbert's energy vector, the speaker [*die Referentin*, the feminine form of the word] stated the following general theorems [...],"[86] and Klein at the 26 July 1918 session of the Royal Scientific Society in Göttingen presented a communication by Noether that bears the same title concerning the invariants of systems of equations that derive from a Lagrangian, which is further testimony to the importance he attributed to Noether's results and to her collaboration. The *Invariante Variationsprobleme* [1918c] would appear in the *Göttinger Nachrichten*[87] with the mention, "the definitive version of the manuscript was prepared only at the end of September." Noether published her own summary of the article in the *Jahrbuch über die Fortschritte der Mathematik*, a yearly collection of abstracts that was the ancestor of the *Zentralblatt* and of *Mathematical Reviews*, now *MathSciNet*. This summary consists of a statement of the two theorems and bears the same title as that article.[88]

Noether submitted the *Invariante Variationsprobleme* to the university with the support of Hilbert and Klein to obtain a habilitation which was awarded in 1919, after the war,[89] after the proclamation of the Weimar Republic and a favorable de-

---

[84] This passage is quoted by Rowe [1999], p. 218, in a different translation.

[85] See Appendix V, p. 167.

[86] "Im Zusammenhang mit der Untersuchungen über den Hilbertschen Energievektor hat die Referentin folgende allgemeine Sätze aufgestellt [...]," *Jahresbericht der Deutschen Mathematiker-Vereinigung*, 27, Part 2 (1918), p. 47. See Dick [1970], p. 15, and [1981], p. 33, and Rowe [1999], p. 221.

[87] For the digitalized version of the *Göttinger Nachrichten*, see, *supra*, note 77.

[88] *Jahrbuch über die Fortschritte der Mathematik*, 46 (1916–1918), p. 770, section Analysis, chapter Calculus of variations.

[89] One can infer from the list of Hilbert's students in his *Gesammelte Abhandlungen*, vol. 3, p. 433, what would have seemed likely, that the war had largely interrupted the presence at the university of the male students, and also delayed the research of those who returned after the war, because none defended theses between 21 December 1914 and 5 June 1918, while the next thesis defense took place 7 July 1920. Judging from the list in Klein's *Gesammelte mathematische Abhandlungen*, vol. 3, pp. 11–13, none of his students defended his thesis during the war, but of course for a different reason. Klein retired in 1913 and had not directed doctoral students since 1911.

cision of the new "Ministry of Science, Arts and Education," and long after the strange incident immortalized in a well-known story about Hilbert's unsuccessful attempt to convince his colleagues to make an exception to the rules barring women from obtaining a habilitation, the first step toward an appointment to the faculty.[90] After having sketched the contents of her earlier publications, she gave the following summary of the article that she had submitted for her habilitation:

> The last two studies that we shall mention concern the differential invariants and the variational problems and are, in part, the result of the assistance that I provided to Klein and Hilbert in their work on Einstein's general theory of relativity. [...] The second study, *Invariante Variationsprobleme*, which I have chosen to present for my habilitation thesis, deals with arbitrary, continuous groups, finite or infinite, in the sense of Lie, and derives the consequences of the invariance of a variational problem under such a group. These general results contain, as particular cases, the known theorems concerning first integrals in mechanics and, in addition, the conservation theorems and the identities among the field equations in relativity theory, while, on the other hand, the converse of these theorems is also given [...].[91]

In the list of habilitations in the 1919 volume of the *Jahresbericht der Deutschen Mathematiker-Vereinigung* we find, "Miss Dr. Emmy Noether has been awarded a habilitation as a *Privatdozentin* in mathematics at the University of Göttingen."[92]

Noether returned to the theory of invariants, though this time they were algebraic invariants, in a paper delivered before the Göttingen Mathematical Society, 5 November 1918, on the invariants of binary forms,[93] and a year later she submitted an article on this subject [1919].

In 1922, there appeared volume III.3 of the *Encyklopädie der mathematischen Wissenschaften*,[94] which was devoted to differential geometry and contained a

---

[90] See the detailed study by Cordula Tollmien [1991] and the article by Tilman Sauer [1999].

[91] "Schließlich sind noch zwei Arbeiten über Differentialinvarianten und Variationsprobleme zu nennen, die dadurch mitveranlaßt sind, daß ich die Herren Klein und Hilbert bei ihrer Beschäftigung mit der Einsteinschen allgemeinen Relativitätstheorie unterstützte. [...] Die zweite Arbeit ,*Invariante Variationsprobleme*', die ich als Habilitationsschrift bezeichnet hatte, beschäftigt sich mit beliebigen endlichen oder unendlichen kontinuierlichen Gruppen, im Lieschen Sinne und zieht die Folgerungen aus der Invarianz eines Variationsproblems gegenüber einer solchen Gruppe. In den allgemeinen Resultaten sind als Spezialfälle die in der Mechanik bekannten Sätze über erste Integrale, die Erhaltungssätze und die in der Relativitätstheorie auftretenden Abhängigkeiten zwischen den Feldgleichungen enthalten, während andererseits auch die Umkehrung dieser Sätze gegeben wird." This text is an extract from the curriculum vitae (*Lebenslauf*) accompanying her habilitation. The original, manuscript, German text is transcribed in Dick [1970], p. 16. It was translated into English in Dick [1981], p. 36, and, with some inaccuracies, in Kimberling [1981], p. 15.

[92] "Fräulein Dr. Emmy Noether hat sich als Privatdozentin der Mathematik an der Universität Göttingen habilitiert." *Jahresbericht der Deutschen Mathematiker-Vereinigung*, 28, Part 2 (1919), p. 36. We note the feminine title, *Privatdozentin*. Appointment as a *Privatdozent* was equivalent to appointment as an assistant professor, but that position implied no remuneration by the university, rather direct remuneration by the students.

[93] *Jahresbericht der Deutschen Mathematiker-Vereinigung*, 28, Part 2 (1918–1919), p. 29.

[94] This encyclopedia had been launched in 1898 under Klein's direction. It was translated into French and published by Gauthier-Villars under the title, *Encyclopédie des sciences mathématiques*

section 10, also designated by III E 1, "New work in the theory of algebraic invariants. Differential invariants" ("Neuere Arbeiten der algebraischen Invariantentheorie. Differentialinvarianten"), written by Weitzenböck and completed in March 1921. In subsection 7, "Differential invariants of infinite groups," he wrote (p. 36), "Recently, differential invariants of infinite groups in connection with a variational principle were considered by E. Noether, using a somewhat more general type of group," [95] and he referred to subsection 27 (*sic* for 28) of that section. In the second part, "Differentialinvarianten," Section C, "Theorie der Differentialformen," this last subsection (no. 28, pp. 68–71) is entitled "Formal calculus of variations and differential invariants" ("Formale Variationsrechnung und Differentialinvarianten") and contains the footnote, "Diese Nr. rührt von E. Noether her," literally "This subsection originates from E. Noether," and was understood after her death as meaning, "This subsection was contributed by E. Noether." Even though her name appears neither in the table of contents (pp. 1–2), nor in the bibliography on page 3, and although it is written in the third person ("E. Noether shows that ..."), this two-page subsection was included in the list of Noether's publications which appeared at the end of her eulogy by van der Waerden.[96] It was subsequently included by Auguste Dick[97] in her bibliography of Noether's writings [1970] [1981] and was reprinted in Noether's *Gesammelte Abhandlungen / Collected Papers*, probably in both cases on the basis of van der Waerden's testimony. In the fifteen-line final paragraph of this short summary, we find references to earlier work that is also cited in the *Invariante Variationsprobleme*, with an additional reference to Klein's paper [1918a], then a restatement of her two theorems which had been published three years earlier:

> The fundamental version of E. Noether shows that to the invariance of J under a group G_ρ (a finite group with ρ essential parameters), there correspond ρ linearly independent divergences; to the invariance under an infinite group which contains ρ arbitrary functions and their derivatives up to order σ, there correspond ρ identities between the Lagrangian expressions and their derivatives up to order σ. In both cases, the converse is valid.[98]

At the end of the above paragraph there is a summary of section 5 of Noether's article: "Given the fact that the Lagrangian expressions are (relative) invariants of the group, one also has a process that generates invariants."[99] This subsection, which

---

pures et appliquées, as the volumes appeared in Germany but, because of the war, the translation was interrupted after 1916, which is to say, before the publication of vol. III.3.

[95] "Neuerdings wurden von E. Noether unter Verwendung eines etwas allgemeineren Gruppenbegriffes Differentialinvarianten von unendlichen Gruppen im Zussammenhang mit einem Variationsprinzip betrachtet," p. 36.

[96] See, *infra*, p. 78. This list appears on p. 475 of van der Waerden [1935].

[97] Dick (1910–1993) held a doctorate in mathematics from the University of Vienna and taught in a high school. She published a book and several articles on Noether, and collaborated in the edition of the works of Erwin Schrödinger (1984).

[98] "Die prinzipielle Fassung bei *E. Noether* zeigt, daß der Invarianz gegenüber einer unendlichen Gruppe, die ρ willkürliche Funktionen bis zur σ^{ten} Ableitung enthält, entsprechen ρ Abhängigkeiten zwischen den Lagrangeschen Ausdrücken und ihren Ableitungen bis zur σ^{ten} Ordnung. In beiden Fällen gilt die Umkehrung," *Encyclopädie*, III.3, p. 71.

[99] "Da die Lagrangeschen Ausdrücke (relative) Invarianten der Gruppe werden, hat man zugleich einen Invarianten erzeugenden Prozeß," *ibid*.

is indeed in Noether's style, may have been written by Noether herself, but this is not entirely clear. In any case, apart from the reference contained in this subsection, we have not found any mention of the *Invariante Variationsprobleme* article in any of Noether's subsequent published works. She did not direct the research of any of her doctoral students toward topics related to variational problems.[100] That suggests that, after having submitted it for her habilitation thesis, she no longer attached great importance to its results.

In Leipzig in 1922, on the occasion of the annual meeting of the German Mathematical Society, she delivered a survey of "Algebraic and Differential Invariants" ("Algebraische und Differentialinvarianten"),[101] and she treated these questions for the last time in her career in an article with the same title in the *Jahresbericht der Deutschen Mathematiker-Vereinigung* [1923]. In the beginning of this paper[102] she remarked that the "naïve and formal" period of research on algebraic invariants had concluded with Hilbert and his utilization of arithmetic methods in algebra, and that, for differential invariants, "this critical period is characterized [...] by the name of Riemann, or, more concretely, [...] by the methods of the formal calculus of variations,"[103] but she cited among her previous works only the articles of 1915 on the existence of rational bases, of 1916 on the existence of a finite basis of invariants for finite groups [1916a], of 1918 on the invariants of differential equations [1918b], and of 1919 on the invariants of binary forms, omitting the *Invariante Variationsprobleme*. On the last page,[104] she referred to Weyl, and to Schouten, whose later papers deal with differential concomitants.

While, as we observed above, Noether never again mentioned her results of 1918 on the variational calculus in print after the 1922 encyclopedia article, if she is indeed its author, she had one occasion to cite the *Invariante Variationsprobleme* when she urged the rejection of a poorly written manuscript submitted to the *Mathematische Annalen* by the physicist Gawrillov Rashko Zaycoff[105] that reproduced and claimed to generalize her results. In a letter of 10 January 1926, written from Blaricum, a village in North Holland,[106] to Einstein,[107] who had evidently asked her

---

[100] See the list of the doctoral theses she directed in Dick [1970], p. 42, and [1981], pp. 185–186.

[101] See Dick [1970], p. 10, and [1981], p. 20.

[102] Noether [1923], p. 177, *Abhandlungen*, p. 436.

[103] "Und diese kritische Periode ist für die algebraischen Invarianten charakterisiert durch den Namen Riemann—oder in sachlicher Hinsicht: [...] durch die Methoden der formalen Variationsrechnung," *ibid.*

[104] Noether [1923], p. 184, *Abhandlungen*, p. 443.

[105] G. R. Zaycoff (1901–1982) studied in Sofia, Göttingen and Berlin, and published articles on relativity and on quantum mechanics. From 1935 on, he worked as a statistician at the University of Sofia.

[106] Blaricum was the residence of the intuitionist mathematician Luitzen Egbertus Brouwer (1881–1966) whom Noether had come to visit for a month in the middle of December 1925 (Alexandrov [1979], cited by Roquette [2008], p. 292). It was also from Blaricum that Weitzenböck had dated the preface of his book [1923].

[107] See the reproduction of this letter, its transcription and a translation in Appendix IV, pp. 161–165.

to evaluate the paper, she justifies her recommendation to reject the article on the grounds that:

> It is first of all a restatement that is not at all clear of the principal theorems of my "Invariante Variationsprobleme" (Göttinger Nachrichten, 1918 or 19), with a slight generalization—the invariance of the integral up to a divergence term—which can actually already be found in Bessel-Hagen (Math. Annalen, around 1922).[108]

Obviously she clearly remembered her work—but not its exact date of publication—and was well aware of Bessel-Hagen's. In the next paragraph she points out that the credit for this generalization is due to Bessel-Hagen and adds a disclaimer that highlights her honesty and lack of ambition: "citing me here [in Zaycoff's second paragraph] is an error" ("daß er mich hier zitiert ist irrtümlich"). After criticizing the nearly incomprehensible computations contained in this paper, she concludes that it does not represent real progress, while her own intent in writing her article had been "to state in a rigorous fashion the significance of the principle and, above all, to state the converse which does not appear here."[109] Then she suggests that a part of the paper might be suitable for some physics journal, and she further suggests that a reference could be made to the statement of her theorems in "Courant–Hilbert," i.e., the recently published book of Courant and Hilbert [1924].[110] Thus, in her own modest way, Noether was conscious of the value of her work. The abstract, rigorous and general point of view that is the mark of all her mathematics is evident in her words, "to state in a rigorous fashion the significance of the principle."

The Klein Jubilee — Noether's correspondence shows great respect for Klein and she dedicated the *Invariante Variationsprobleme* to him on the occasion of his academic jubilee.[111] It used to be a frequent practice in German universities to celebrate the fiftieth anniversary of an eminent professor's doctorate, *das goldene Doktorjubiläum*. In 1916 Hilbert had written an article for the jubilee of Hermann Amandus Schwarz, which was reprinted in the same volume of the *Göttinger Nachrichten* as Noether's [1916b]. Max Noether's jubilee was celebrated 5 March 1918. Klein's doctorate having been awarded 12 December 1868 at the University of Bonn, his academic jubilee was celebrated in Göttingen, at the university on 10 December 1918, and at the Mathematical Society two days later with a lecture on his scientific work delivered by Paul Koebe.[112]

---

[108] "Es handelt sich zuerst um eine nicht allzu durchsichtige Wiedergabe der Hauptsätze meiner 'Invarianten Variationsprobleme' (Göttinger Nachrichten 1918 oder 19), mit einer geringen Erweiterung—Invarianz des Integrals bis auf Divergenzglied—die sich schon bei Bessel-Hagen findet (Math. Annalen etwa 1922)." It was Bessel-Hagen's article [1921], analyzed below, in Chap. 4, p. 91, that formally introduced the symmetries up to divergence.

[109] "Mir kam es in den 'Invarianten Variationsproblemen' nur auf die scharfe Formulierung der Tragweite des Prinzips an, und vor allem auf die Umkehrung[,] die hier nicht herein spielt."

[110] See, *infra*, Chap. 4, p. 95.

[111] Jubilee, from the Hebrew *yovel*, horn, which became a metonymy for a fiftieth year because, in the biblical calendrical cycle, every fiftieth year was to be inaugurated by sounding such a horn.

[112] "On the scientific work of F. Klein, in particular on the theory of automorphic functions" ("Über F. Kleins wissenschaftliche Arbeiten, insbesondere die die Theorie der automorphen Funktionen

1.5 After Göttingen

After the period 1915–1918, Noether directed her research toward abstract algebra, the theory of ideals and the representation theory of algebras, and became one of the most important mathematicians of her time. She was deprived of her employment by the Nazis and compelled to leave Göttingen where her forceful personality and great talent had attracted many students. Since she was not a civil servant, she could not be dismissed directly, but she was put on leave with full pay on 25 April 1933.[113] On 13 September, paragraph 3 of the law of 7 April 1933 that excluded all persons of non-aryan descent from the civil service was applied to all the Jews who taught in the universities, civil servants or not, with few exceptions. On that day, she wrote to Richard Brauer:

> Since presently paragraph 3 comes into effect—I was notified today that my permission to teach has been rescinded in accordance with this paragraph [. . .].[114]

In the next lines she asks whether Brauer has any prospect of employment, then discusses her own possibilities for the coming academic year, and, in the last part of her letter, she gives news of the mathematical results of three young Göttingen mathematicians, among whom Max Deuring, who had defended in 1931 his doctoral thesis written under her direction, and Ernst Witt, who had joined the Nazi party in May and defended his thesis in July. Noether left Göttingen shortly thereafter, visited Russia briefly, but preferred refuge in the United States where, until her premature death in 1935, she taught at Bryn Mawr College[115] outside Philadelphia, a women's undergraduate school with a small graduate school to which a number of male students had been admitted since 1931. She also participated very actively in the mathematical life of the Institute for Advanced Study at Princeton, a short train ride from Philadelphia. After her death she was replaced at Bryn Mawr by Nathan Jacobson for the 1935–1936 academic year.[116] Numerous articles and books have discussed her life and her work as an algebraist.[117]

---

betreffenden"). See *Jahresbericht der Deutschen Mathematiker-Vereinigung*, 28, Part 2 (1919), p. 30. This ceremony is mentioned in the preface to Klein's *Gesammelte mathematische Abhandlungen*, vol. 1, p. iii, and the (unsigned) text of an address delivered on that occasion is printed on the pages that follow the preface. (We observe that the editors of the *Gesammelte mathematische Abhandlungen*, vol. 1 (1920), acknowledged (p. v) the assistance of Miss E. Noether for the correction of the proofs.) Also see, in the *Jahresbericht der Deutschen Mathematiker-Vereinigung*, 27, Part 2 (1918), pp. 59–60, a letter of congratulations from the DMV, and on p. 63, the announcement of the formation of a foundation by Klein's friends and students and another letter of congratulations by the proponents of this foundation, which was delivered by Robert Fricke.

[113] Segal [2003], p. 125.

[114] "Da augenblicklich §3 in Aktion tritt—ich habe heute die Mitteilung der entzogenen Lesebefügnis nach diesem [. . .]," letter from Emmy Noether to Richard Brauer, in the Bryn Mawr archives, partially translated in Curtis [1999], pp. 213–214.

[115] Some of her German mathematics books can still be found in the mathematics department.

[116] *Notices of the American Mathematical Society*, October 2000, p. 1061. Jacobson (1910–1999) had attended her lectures in Princeton. He later was the editor of her *Collected Papers*.

[117] See Dick [1970] [1981], Kimberling [1981], Srinivasan and Sally [1983], Teicher [1999], Curtis [1999], etc.

Chapter 2
The Noether Theorems

This chapter will deal briefly with the results stated and proved by Noether in the *Invariante Variationsprobleme*[1] [1918c]. Her originality in this article consisted in dealing with problems that arose either in classical mechanics (the first theorem) or in general relativity (the second theorem). We emphasize what has been ignored by most authors who have cited this article, that in it Noether treated a problem of very great generality, since she dealt with a Lagrangian of arbitrary order with an arbitrary number of independent variables,[2] as well as an arbitrary number of dependent variables, and considered the invariance of such Lagrangians under the action of "groups of infinitesimal transformations." The infinitesimal transformations in question that form, in modern mathematical terminology, Lie algebras of finite dimension, ρ, or of infinite dimension are genuine generalizations of the usual vector fields since their components depend not only on the independent and dependent variables, as is the case for the infinitesimal generators of Lie groups of transformations, but also on the successive derivatives of the dependent variables. In other words, the infinitesimal symmetries that she considered might depend not only on the field variables but also on their derivatives of order 1 or higher.[3]

---

[1] For a more mathematical discussion, see Olver [1986a], as well as the numerous references cited therein, or see Kosmann-Schwarzbach [1985] and [1987].

[2] This theory has since been developed by numerous authors (see, *infra*, Chap. 7). Leonid Dickey [1991] [1994] calls it the "multi-time" theory if there are more than one independent variable. Other authors reserve that description for the case in which several independent variables play a role analogous to that of time among the four variables of relativistic space-time. Recall that variational problems with the time as the single independent variable correspond to problems in mechanics, while those with several independent variables arise in field theory. In nonrelativistic field theory, the three independent variables represent coordinates in space while, in relativistic field theory, the four independent variables represent coordinates in space-time.

[3] Recall that Noether uses the term "invariance" rather than the terms "symmetry" or "symmetry transformation" which have now become standard. She distinguishes between global transformations that form a "continuous group in the sense of Lie" and the infinitesimal transformations that are the generators of the one-parameter subgroups of such a group, while the vanishing of a Lie derivative is expressed by "Lie's differential equation." We often abbreviate "infinitesimal symmetry" as "symmetry."

Y. Kosmann-Schwarzbach, *The Noether Theorems*, Sources and Studies in the History of Mathematics and Physical Sciences, DOI 10.1007/978-0-387-87868-3_3,

2.1 Preliminaries

Let us recall that a conservation law in mechanics is a quantity that depends on the configuration variables and their derivatives, and which remains constant during the motion of the system. Therefore, a conservation law is also called a first integral of the equation of motion. In a field theory described by an evolution equation of the form $\dfrac{\partial u}{\partial t} = F(x, u, u_x, \ldots)$, a conservation law is a relation of the form $\dfrac{\partial T}{\partial t} + \sum_{i=1}^{n-1} \dfrac{\partial A_i}{\partial x_i} = 0$, where the $x = (x_1, \ldots, x_{n-1})$ are the space variables and $t = x_n$ is time, and where $A_1, \ldots, A_{n-1}$ and T are functions of the independent variables, and of the field variables u and their derivatives with respect to the space variables, which relation is satisfied when the field equations are satisfied. In physics a conservation law is also called a continuity equation. If the conditions for the vanishing of the quantities being considered at the boundary of a domain of the space variables, $x_1, \ldots, x_{n-1}$, are satisfied, one deduces, by an application of Stokes's theorem,[4] that the integral of T over this domain is constant over the course of time. One then says that T is the density of a conserved quantity. More generally, in the presence of several variables, when no single one representing time is distinguished from the others, conservation laws define integrals which depend exclusively on the boundary of the domain of integration. In particular, in the case of two independent variables, one obtains line integrals which depend exclusively on the endpoints of the path under consideration.[5]

In the short introduction to her article, Noether cites, in the text or in the notes, the earlier work of Hamel [1904a, b], Herglotz [1911], Lorentz[6] and his student Fokker [1917], Weyl,[7] Klein [1918b] and Kneser [1918].[8] She explains that her work is based on "a combination of the methods of the formal calculus of variations and Lie's theory of groups"[9] and also that there is a close relation between her work and Klein's [1918b].

---

[4] Stokes's theorem, also called the Gauss–Ostrogradsky theorem or formula, states that the integral of an exact form $d\beta$ over a domain Ω is equal to the integral of β on the boundary of Ω; in particular, the integral of the divergence of a vector field on a domain is equal to an integral on the boundary of that domain. This result was due to George Gabriel Stokes (1819–1903), and eventually the general formula came to bear his name.

[5] This is the type of conservation law that is to be found in continuum mechanics and particularly in elasticity theory. See, *infra*, p. 147.

[6] Lorentz's articles [1915] [1916] on Einstein's theory of gravitation appeared between 1915 and 1917.

[7] She was probably referring to the article that Weyl had submitted to the *Annalen der Physik* on 8 August 1917 (Weyl [1917]), and maybe also to "Zeit, Raum, Materie" (Weyl [1918b]), which Klein had cited in his note [1918b].

[8] See Chap. 1, pp. 35–37 and 39, for comments on the work of Noether's predecessors that she cites here.

[9] See, *infra*, Chap. 7, p. 138, considerations of modern developments of the first of these two theories. The theory of Lie groups is now a vast domain of pure mathematics and an indispensable tool in modern physics.

2.2 The First Theorem: Conservation Laws

Noether's first theorem, a generalization of several conservation theorems that were already known in mechanics, and its converse are the subject of the first part of Section 2 and of Section 3. She considers a multiple integral,

$$I = \int \cdots \int f\left(x, u, \frac{\partial u}{\partial x}, \frac{\partial^2 u}{\partial x^2}, \cdots\right) dx,$$

of a higher-order Lagrangian f that is a function of n independent variables, $x_1, \ldots, x_\lambda, \ldots, x_n$, and of μ dependent variables, $u_1, \ldots, u_i, \ldots, u_\mu$, as well as of their derivatives up to a fixed but arbitrary order, κ. She then considers a variation of u, $\delta u = (\delta u_i)$, and derives identity (3),

$$\sum_{i=1}^{\mu} \psi_i \, \delta u_i = \delta f + \text{Div} \, A,$$

where the ψ_i are the Lagrangian expressions, which is to say the components of the variational derivative (Euler–Lagrange derivative) of f, and where the components A_λ of A are linear in the variation δu and in its derivatives. The opposite of the quantity A is now called the Legendre transform of the Lagrangian f. Here Div is the ordinary divergence, $\text{Div} \, A = \sum_{\lambda=1}^{n} \dfrac{\partial A_\lambda}{\partial x_\lambda}$, of $A = (A_1, \ldots, A_n)$ considered as a vector in n-dimensional space, and δf is the variation of f corresponding to the variation δu of u, while the variation of x is assumed to vanish.[10] Identity (3) is obtained by an integration by parts. In the case where $n = 1$, the case of a simple integral, Noether gives an expression for A for an arbitrary μ, first for $\kappa = 1$, which yields what she calls Heun's[11] "central Lagrangian equation," then for an arbitrary κ,[12] and then she states her theorem:

I. *If the integral I is invariant under a [group] $\mathfrak{G}_\rho$, then there are ρ linearly independent combinations among the Lagrangian expressions which become divergences— and conversely, this implies the invariance of I under a [group] $\mathfrak{G}_\rho$. The theorem remains valid in the limiting case of an infinite number of parameters.*

---

[10] Because of the geometric interpretation of these quantities in terms of vector bundles over manifolds, we have called such a variation a "vertical vector field" (Kosmann-Schwarzbach [1980] [1985] [1987]). Olver [1986a] calls it an "evolution vector field." The variation δf of f is both the Lie derivative of f in the direction of the vector field δ and the action of the prolongation of δ on f.

[11] Karl Heun (1859–1929), who defended his thesis in Göttingen in 1881, was the author of numerous articles and several books on differential equations and their applications to mechanics. He held the chair of mechanics at the University of Karlsruhe from 1902 to 1923. Georg Hamel was his assistant from 1902 to 1905, later Fritz Noether, the second of Emmy's three brothers, became his assistant from 1909 to 1917 and it was he who wrote the eulogy of Heun that appeared in the *Zeitschrift für angewandte Mathematik und Mechanik*.

[12] This formula can be read on Noether's postcard of 15 February 1918 to Klein, reproduced in Appendix I, p. 149. See, *supra*, Chap. 1, p. 46.

Noether explains that "in the one-dimensional case," that is, when $n = 1$, one obtains first integrals, while, "in higher dimensions," i.e, when $n > 1$, "one obtains the divergence equations which, recently, have often been referred to as *conservation laws*." By the "limiting case" included in the statement of Theorem I is meant the case in which the elements of the group depend on an infinite but denumerable set of parameters, as opposed to the case dealt with in her Theorem II.

The very short proof is in Section 2. Noether assumes that the action integral I is invariant. Actually, she assumes a more restrictive hypothesis,[13] the invariance of the integrand, $f\,dx$, which is to say $\delta(f\,dx) = 0$. This hypothesis is expressed by relation (11),

$$\bar{\delta}f + \mathrm{Div}(f.\,\Delta x) = 0,$$

on which the proof depends. Here $\bar{\delta}f$ is the variation of f for the variation

$$\bar{\delta}u_i = \Delta u_i - \sum \frac{\partial u_i}{\partial x_\lambda}\Delta x_\lambda.$$

In fact, this is how Noether introduced the components of the "vertical generalized vector field" $\bar{\delta}$ associated with δ, and $\bar{\delta}f$ is the Lie derivative of f in the direction of the vertical vector field $\bar{\delta}$.[14] Noether then uses identity (3),

$$\sum \psi_i \bar{\delta}u_i = \bar{\delta}f + \mathrm{Div}\,A,$$

which, in view of the invariance hypothesis (11), can be written

$$\sum \psi_i \bar{\delta}u_i = \mathrm{Div}\,B, \quad \text{with } B = A - f.\,\Delta x,$$

which is the fundamental relation (12) in her article.[15]

---

[13] This hypothesis is sufficient but not necessary for I to be invariant, as we shall see in Section 4.1 of Chap. 4 concerning symmetries up to divergence.

[14] Kneser [1918] used the variation $\delta x - \dot{x}\delta t$ for the integral $\int H(x,\dot{x})\,dt$ in his study of mechanics governed by a Lagrangian. One can find a formula analogous to Noether's expression for $\bar{\delta}u_i$ in Weyl [1918b], p. 186 (3rd ed., 1918, IV.28, p. 201; 5th ed., 1923, IV.30, p. 234; English translation, 1922, p. 234), and already in Weyl [1917], p. 123 (*Gesammelte Abhandlungen*, vol. 1, p. 676), we find the following variation of the coefficients g^{ik} of the space-time metric of general relativity,

$$\Delta g^{ik} = \delta g^{ik} - \varepsilon \frac{\partial g^{ik}}{\partial x_a}\xi_a,$$

which he calls a "virtual displacement" (*virtuelle Verrückung*). Rosenfeld ([1930], p. 117) calls it the "substantial variation" (*substantielle Variation*). The use of a "vertical representative" has now become standard. See Boyer [1967], and Bluman and Kumei [1989].

[15] In the notation of Kosmann-Schwarzbach [1985], the hypothesis may be written $X \cdot L = 0$, and the fundamental relation may be written $\langle EL, \tilde{X} \rangle = -d_M(\mathscr{F}L \circ \tilde{X} + i_{X_M}L)$, the notation $\mathscr{F}L$ recalling that this operator is the Legendre transform which is also called the fiber derivative of the Lagrangian L. Here $\tilde{X}$ denotes the vertical representative of the vector field X, and X_M is its projection onto the space of independent variables, EL is the Euler–Lagrange derivative of the Lagrangian L, and d_M corresponds to the divergence operator with respect to the independent variables.

It follows directly from this relation that if the Euler–Lagrange equations, $\psi_i = 0$, are satisfied, then B is a conservation law. Noether has thus proved that to each infinitesimal invariance transformation of a Lagrangian, i.e., to each pair $(\Delta u, \Delta x)$ satisfying relation (11), there corresponds a linear combination of the Lagrangian expressions which is a divergence. She then shows that the conservation laws associated with ρ linearly independent infinitesimal invariance transformations are themselves linearly independent, provided that a certain class of infinitesimal invariance transformations be excluded.[16] Actually, although the rigor of Noether's proofs does not conform to current standards, she does not neglect to point out difficulties that would be explained in terms of equivalence relations in the mathematical work of the 1970s and 1980s.[17]

Then, in Section 3, Noether proves the converse of her first theorem: if ρ linearly independent relations among the Lagrangian expressions are divergences, then there exists a ρ-parameter family of linearly independent infinitesimal invariance transformations, and thus the variational integral is invariant under the action of a ρ-parameter continuous group. She then observes the complications that result from having considered variations which depend on the derivatives of the dependent variables which are of two types: on the one hand, the above result is only valid after passing to equivalence classes of symmetries and of conservation laws, and, on the other, the integration of a vector field $\bar{\delta}$ in general requires the solution of a system of partial differential equations, while the flow of a vector field in the usual sense is determined by the integration of a system of ordinary differential equations. She further observes that a study of the infinitesimal symmetries of an equivalent Lagrangian, i.e., differing only by a divergence and thus possessing the same Lagrangian expressions, necessarily leads to the introduction of generalized vector fields.[18] In terms of the symmetries up to divergence that would be introduced a few years later by Bessel-Hagen,[19] the relationship between the infinitesimal symmetries of two Lagrangians that differ only by a divergence can be stated very simply: any symmetry up to divergence of the one is a symmetry up to divergence of the other.

In Section 5, Noether studies the action of an invariance transformation on a conservation law. She first determines in which case one can assume the symmetry to be a vertical vector field, and in that case she concludes that $\operatorname{Div} B$ is relatively invariant. Restricting herself to the case of a single independent variable, she shows that, under the action of the invariance group, first integrals remain first integrals. Since, in its infinitesimal version, relative invariance amounts to equality to 0 modulo a divergence, this result implies that under the action of an infinitesimal symmetry of the variational problem, each conservation law is transformed into another conservation law.[20] Noether also considers in this section the consequences of the second

---

[16] P. 242, note 1 (p. 9, note 13, in the above translation).

[17] See Vinogradov [1984a], Olver [1986a], and Krasil'shchik and Vinogradov [1997].

[18] P. 245 (p. 12, in the above translation).

[19] See, *infra*, Chap. 4, p. 91.

[20] This subject has been further developed by Ibragimov [1983] and by Olver [1986a]. See also Benyounès [1987].

theorem which we will now discuss. Her note on p. 251[21] poses the question whether the invariance up to divergence of the Euler–Lagrange equations implies that of the Lagrangian itself. She shows that such is the case only under restrictive conditions, and she refers to Engel [1916] who had treated the case of a first-order Lagrangian with a single independent variable.

2.3 The Second Theorem: Differential Identities

In her first section, Noether also states a second theorem:[22]

II. *If the integral I is invariant under a [group] $\mathfrak{G}_{\infty\rho}$, depending on [$\rho$] arbitrary functions and their derivatives up to order σ, then there are ρ identities among the Lagrangian expressions and their derivatives up to order σ. Here as well the converse is valid.*

Here the elements of the invariance group depend on functions, each of which can be considered to be a continuous, nondenumerable set of parameters. She explains the precautions that must be taken—in modern terms, the introduction of an equivalence relation on the symmetries—for the converse to be valid. The proof of this theorem is provided in the second part of Section 2, where Noether assumes the existence of ρ symmetries of the Lagrangian, each of which depends linearly on an arbitrary function $p^{(\lambda)}$ ($\lambda = 1, 2, \ldots, \rho$) of the variables $x_1, x_2, \ldots, x_n$, and its derivatives up to order σ. Such a symmetry is defined by a vector-valued linear differential operator of order σ, which we denote by $\mathscr{D}^{(\lambda)}$, and whose components we denote by $\mathscr{D}_i^{(\lambda)}$, $i = 1, 2, \ldots, \mu$, or, in Noether's notation for the right-hand side,

$$\mathscr{D}_i^{(\lambda)}(p^{(\lambda)}) = a_i^{(\lambda)}(x, u, \ldots)p^{(\lambda)}(x) + b_i^{(\lambda)}(x, u, \ldots)\frac{\partial p^{(\lambda)}}{\partial x} + \cdots + c_i^{(\lambda)}(x, u, \ldots)\frac{\partial^\sigma p^{(\lambda)}}{\partial x^\sigma}.$$

Noether then introduces, without giving it a name or a particular notation, the adjoint operator,[23] $(\mathscr{D}_i^{(\lambda)})^*$, of each of the $\mathscr{D}_i^{(\lambda)}$, which, by construction, satisfies

$$\psi_i \mathscr{D}_i^{(\lambda)}(p^{(\lambda)}) = (\mathscr{D}_i^{(\lambda)})^*(\psi_i)p^{(\lambda)} \quad \text{modulo divergences} \quad \text{Div}\, \Gamma_i^{(\lambda)},$$

where the $\Gamma_i^{(\lambda)}$ are linear in the ψ_j ($j = 1, 2, \ldots, \mu$) and their derivatives. Identities (13),

$$\sum \psi_i \bar{\delta} u_i^{(1)} = \text{Div}\, B^{(1)}; \ldots \qquad \sum \psi_i \bar{\delta} u_i^{(\rho)} = \text{Div}\, B^{(\rho)},$$

---

[21] P. 17, note 22, in the above translation.

[22] For a modern study of Noether's second theorem, one should consult Olver [1986b]. An analysis of the historical context of Noether's second theorem and its significance may be found in Brading [2005]. See also Brading and Brown [2003].

[23] For the concept of the adjoint operator, see Volterra [1913].

had already been derived in the first part of this section. In view of identity (3) and the assumption of the invariance of the Lagrangian, they may be written as

$$\sum_{i=1}^{\mu} \psi_i \, \mathscr{D}_i^{(\lambda)}(p^{(\lambda)}) = \operatorname{Div} B^{(\lambda)} \qquad (\lambda = 1, 2, \ldots, \rho).$$

These relations imply

$$\sum_{i=1}^{\mu} (\mathscr{D}_i^{(\lambda)})^* (\psi_i) \, p^{(\lambda)} = \operatorname{Div}(B^{(\lambda)} - \Gamma^{(\lambda)}),$$

where $\Gamma^{(\lambda)} = \sum_{i=1}^{\mu} \Gamma_i^{(\lambda)}$. From this, by an application of Stokes's theorem and the Du Bois-Reymond lemma,[24] it follows that, since the $p^{(\lambda)}$ are arbitrary,

$$\sum_{i=1}^{\mu} (\mathscr{D}_i^{(\lambda)})^* (\psi_i) = 0,$$

for $\lambda = 1, 2, \ldots, \rho$. These are the ρ differential relations among the components ψ_i of the Euler–Lagrange derivative of the Lagrangian f that are identically satisfied. These differential identities are obtained in formula (16) of the article.

It should be observed that, in her postcard of 15 February 1918 to Klein,[25] Noether had already announced this result, but only for the very special case where the variation of the u_i under consideration is $p^{(\kappa)} \dfrac{\partial u_i}{\partial x_\kappa}$, for a fixed κ chosen among the values $1, 2, \ldots, n$, and where $p^{(\kappa)}$ is an arbitrary function of $x_1, \ldots, x_n$, a variation which corresponds to the most general vector field in the direction of the coordinate line x_κ. In this case, quite simply, $\mathscr{D}_i^{(\kappa)}(p^{(\kappa)}) = \dfrac{\partial u_i}{\partial x_\kappa} p^{(\kappa)}$. The differential operator $\mathscr{D}_i^{(\kappa)}$ is of order 0 and thus coincides with its adjoint, and the relation $\sum_{i=1}^{\mu} (\mathscr{D}_i^{(\kappa)})^* (\psi_i) = 0$ reduces to

$$\sum_{i=1}^{\mu} \psi_i \frac{\partial u_i}{\partial x_\kappa} = 0,$$

the result announced at the end of the postcard.

---

[24] This lemma states that, if the integral of the product of a given function by an arbitrary function vanishes, then the given function vanishes. Noether does not cite this result which is attributed to Paul Du Bois-Reymond (1831–1889), a mathematician of Swiss origin who taught at the Universities of Heidelberg and Freiburg, then in Berlin, and who published on the theory of functions and partial differential equations.

[25] See Chap. 1, p. 46, and Appendix I, p. 149.

Noether returns to her second theorem in Section 6 where she observes that her identities (16) may be written

$$\sum_{i=1}^{\mu} a_i^{(\lambda)} \psi_i = \operatorname{Div} \chi^{(\lambda)},$$

where each $\chi^{(\lambda)}$ is defined by a linear differential operator acting on the Lagrangian expressions ψ_i. In fact, in $(\mathcal{D}_i^{(\lambda)})^*(\psi_i)$, the first term is $a_i^{(\lambda)} \psi_i$, and subsequent terms may be written in the form of a divergence. From the fact that identities (16) must imply the divergence relations when the $p^{(\lambda)}$ are constants, and from the linearity of each expression with respect to the ψ_i, she deduces that, in the case of a group depending on arbitrary functions, the divergences $\operatorname{Div} B^{(\lambda)}$ which appear in relations (13) are linear combinations of the $\operatorname{Div} \chi^{(\lambda)}$, that is to say,

$$\operatorname{Div} B^{(\lambda)} = \operatorname{Div} C^{(\lambda)}, \qquad \text{where} \quad C^{(\lambda)} = \sum_{\kappa=1}^{\rho} \alpha_{(\kappa)}^{(\lambda)} \chi^{(\kappa)}.$$

Because the $\chi^{(\kappa)}$ are linear in the ψ_i, each $C^{(\lambda)}$, and not only the divergence of $C^{(\lambda)}$, vanishes once the Euler–Lagrange equations $\psi_i = 0$ are satisfied. Furthermore, from the equality of the divergences of $B^{(\lambda)}$ and $C^{(\lambda)}$, it follows that

$$B^{(\lambda)} = C^{(\lambda)} + D^{(\lambda)}$$

for some $D^{(\lambda)}$ whose divergence vanishes identically, which is to say, independently of the satisfaction of the Euler–Lagrange equations.

Noether called the conservation laws obtained in the case of invariance groups depending on arbitrary functions and their derivatives "improper divergence relations" (*uneigentliche Divergenzrelationen*), a term that has not been retained in the literature of general relativity where those laws play an important role, while she called all other conservation laws "proper" (*eigentliche*). In fact, in general relativity, improper conservation laws or, more precisely, conservation laws that correspond to divergences which vanish identically and are thus satisfied independently of the field equations, are generally called *strong* conservation laws,[26] while the conservation laws obtained from the first theorem are called *weak* laws.[27]

---

[26] See Bergmann [1958], Trautman [1962], Goldberg [1980].

[27] See, *infra*, Chap. 6, p. 126. In [1986a], Olver introduced the concept of *trivial conservation laws*, distinguishing between the *trivial conservation laws of the first kind*, where Div $C = 0$ and the expression C itself, and not only its divergence, vanish when the Euler–Lagrange equations are satisfied, and the *trivial conservation laws of the second kind*, corresponding to the strong laws, where Div D vanishes identically. In this terminology, what Noether calls an improper law is thus a trivial law in the sense of Olver, the sum of a trivial law of the first kind and a trivial law of the second kind. Olver's introduction of an equivalence relation in which two conservation laws are identified if they only differ by a trivial conservation law permits restating Noether's first theorem and its converse with precision: suitably defined equivalence classes of symmetries are in one-to-one correspondence with the equivalence classes of conservation laws. For this, also see Vinogradov [1984a].

The distinction between proper conservation laws, which appear when there is a finite-dimensional Lie group of invariance transformations, and improper conservation laws, which are related to the existence of invariance transformations that depend on arbitrary functions, is due to Noether, who used it in the second part of her sixth and last section, which will be analyzed below, to clarify in a very simple fashion a feature of the law of conservation of energy in general relativity.

In Section 4, Noether shows that the assumption of linearity for the differential operators $\mathscr{D}_i^{(\lambda)}$ is not a restriction, and carefully studies the converse of her second theorem. She adds a long note at the end of that section (pp. 249–250, note 1; pp. 15–16, note 20, in the above translation) in which she shows essentially that if one replaces a Lagrangian by an equivalent Lagrangian, equivalent in the sense that it yields the same Euler–Lagrange equations and thus only differs from the original Lagrangian by the addition of a divergence, then "Δx and Δu will, in general, contain derivatives of the u." That is to say, for a Lagrangian equivalent to a given Lagrangian, the symmetries corresponding to the same conservation law will, in general, be generalized symmetries depending on the dependent variables and their derivatives, even if the symmetries of the given Lagrangian are classical symmetries, that is to say, infinitesimal symmetries in the sense of Lie. (Such symmetries are often called "Lie symmetries.") She then gives the example of a Lagrangian equivalent to Hilbert's Lagrangian,[28] which had been introduced into the general theory of relativity by Einstein [1916a, b] and studied by Klein [1918b], for which she determines the symmetries explicitly. They depend not only on the components of the metric but also on their first and second derivatives as well as on four arbitrary functions and their first derivatives.

2.4 Conclusion: The Discussion of Hilbert's Assertion

In the conclusion of her article, in Section 6, Noether examines Hilbert's assertion[29] that, in the case of general relativity and in that case only, there are no proper conservation laws, and she shows that the situation is better understood "in the more general setting of group theory."[30] In particular, she explains the apparent paradox that arises from the consideration of the finite-dimensional subgroups of groups that depend on arbitrary functions. Those subgroups cannot, under any circumstances, correspond to proper conservation laws and, in particular, "given I invariant under the group of translations, then the energy relations are improper if and only if I is invariant under an infinite group which contains the group of translations as a subgroup." In conclusion, Noether writes,

---

[28] Hilbert's Lagrangian is equal to the scalar curvature multiplied by the square root of the absolute value of the determinant of the space-time metric.

[29] Hilbert's assertion was published in Klein [1918a] which also contained Klein's answer to Hilbert's remarks. See, *infra*, p. 66.

[30] "In verallgemeinerter gruppentheoretische Fassung," p. 254.

As Hilbert expresses his assertion, the lack of a proper law of [conservation of] energy constitutes a characteristic of the "general theory of relativity." For that assertion to be literally valid, it is necessary to understand the term "general relativity" in a wider sense than is usual, and to extend it to the aforementioned groups that depend on n arbitrary functions.

In her final footnote, Noether notes the relevance of Klein's [1910] observation in the spirit of his Erlangen program,[31] which she paraphrases as "the term *relativity* that is used in physics should be replaced by *invariance with respect to a group.*" She thus extrapolates from the problems arising from the invariance group of the equations of mechanics and from that of the equations of general relativity to a general theory of invariance groups of variational problems, distinguishing with clairvoyance the case of invariance groups that are finite-dimensional Lie groups from groups of transformations that depend on arbitrary functions and are therefore essentially infinite-dimensional. This latter case would become, in the work of Weyl and, much later, Chen Ning Yang and Robert L. Mills, gauge theory.

---

[31] On Klein's *Erlanger Programm*, see, *supra*, Introduction, p. 26, note 3.

Chapter 3
The Noether Theorems as Seen by Contemporaries and by Historians of Science

In 1918 in Göttingen, Klein was certainly the most senior and important member of the faculty, Hilbert was a dominant, internationally acknowledged mathematician, while Weyl was a slightly younger contemporary of Noether whose outstanding talent was already recognized. Einstein was a celebrity, then residing in Berlin, whose work was being followed closely in Göttingen. We shall review the way in which Noether's achievements were perceived and acknowledged by each of them at the time, before analyzing later testimonies to her work as a mathematician and the extremely modest role assigned to her early work up to and including the *Invariante Variationsprobleme*.

3.1 References to Noether in the Works of Klein, Hilbert and Weyl, and in Einstein's Correspondence

One should be able to form an idea of the role that Noether played in the development of general relativity theory by looking at several texts in the mathematical publications and correspondence of her contemporaries Klein (1849–1925), Hilbert (1862–1943), Weyl (1885–1955) and Einstein (1879–1955) where they acknowledged, Klein more generously than any of his colleagues, and none of them very frequently, Noether's contribution to their explorations of the implications of that theory. Of particular interest are the notes that Klein added to his 1918 papers when they were re-issued in the first volume of his *Gesammelte mathematische Abhandlungen*. Some of this material has been studied with great competence by Jagdish Mehra [1972], Hans A. Kastrup [1987] and, more recently, David E. Rowe [1999].[1] Several references to Noether's work have already been mentioned briefly in the preceding chapters, but these testimonies bear repetition because of the neglect of her contribution in the research of the next generation of physicists that will be demonstrated in the following chapters.

---

[1] Also see Pais [1982] and Byers [1999].

Y. Kosmann-Schwarzbach, *The Noether Theorems*, Sources and Studies in the History
of Mathematics and Physical Sciences, DOI 10.1007/978-0-387-87868-3_4,
© Springer Science+Business Media, LLC 2011

Klein and Hilbert — We have cited above (Chap. 1, p. 45, note 78), Hilbert's letter to Einstein of 27 May 1916. A year and a half later, there appeared "On Hilbert's first note on the foundations of physics," an article in three parts by Klein [1918a] in which he published extracts from a letter he had writtten to Hilbert, followed by extracts from Hilbert's answer, then by his own additional remarks. In his letter, after seven pages of mathematics, Klein declares, "Here, I have to make an essential parenthetical statement," and continues,

> You know that Miss Noether advises me continually regarding my work, and that in fact it is only thanks to her that I have understood these questions. When I was speaking recently to Miss Noether about my result concerning your energy vector, she was able to inform me that she had derived the same result on the basis of developments of your note (and thus not from the simplified calculations of my section 4) more than a year ago, and that she had then put all of that in a manuscript (which I was subsequently able to read). She simply did not set it out as forcefully as I recently did at the Mathematical Society (22 January [1918]).[2]

Hilbert's answer begins with a remark about Noether which is very clear:

> I fully agree in fact with your statements on the energy theorems: Emmy Noether, on whom I have called for assistance more than a year ago to clarify this type of analytical questions concerning my energy theorem, found at that time that the energy components that I had proposed—as well as those of Einstein—could be formally transformed, using the Lagrange differential equations (4) and (5) of my first note, into expressions whose divergence vanishes *identically*, that is to say, without using the Lagrange equations (4) and (5).[3]

One would say in a more modern language that the conservation laws are valid "off-shell." Hilbert then states a conjecture, the assertion (*Behauptung*) which Noether will elucidate in the last section of her article:

> Indeed I believe that in the case of general relativity, i.e., in the case of the *general* invariance of the Hamiltonian function, the energy equations which in your opinion correspond

---

[2] "Hier habe ich eine wesentliche Einschaltung zu machen. Sie wissen, daß mich Frl. Nöther bei meinen Arbeiten fortgesetzt berät and daß ich eigentlich nur durch sie in die vorliegende Materie eingedrungen bin. Als ich nun Frl. Nöther letzthin von meinem Ergebnis betr. Ihren Energievektor sprach, konnte sie mir mitteilen, daß sie dasselbe aus den Entwicklungen Ihrer Note (also nicht aus den vereinfachten Rechnungen meiner Nr. 4) schon vor Jahresfrist abgeleitet und damals in einem Manuskrit festgelegt habe (in welches ich dann Einsicht nahm); sie hatte es nur nicht mit solcher Entschiedenheit zur Geltung gebracht, wie ich kürzlich in der Mathematischen Gesellschaft (22. Januar)," Klein [1918a], p. 476, and *Gesammelte mathematische Abhandlungen*, vol. 1, p. 559. Parts of this passage and the next have been quoted in English translation by Mehra [1974], p. 70, note 129a, Rowe [1999], pp. 213–214, and Pais [1982], p. 276.

[3] "Mit Ihren Ausführungen über den Energiesatz stimme ich sachlich völlig überein: Emmy Noether, deren Hilfe ich zur Klärung derartiger analytischer meinen Energiesatz betreffenden Fragen vor mehr als Jahresfrist anrief, fand damals, daß die von mir aufgestellten Energiekomponenten—ebenso wie die Einsteinschen—formal mittels der Lagrangeschen Differentialgleichungen (4), (5) in meiner ersten Mitteilung in Ausdrücke verwandelt werden können, deren Divergenz *identisch*, d. h. ohne Benutzung der Lagrangeschen Gleichungen (4), (5) verschwindet," Klein [1918a], p. 477, and *Gesammelte mathematische Abhandlungen*, vol. 1, pp. 560–561.

to the energy equations of the theory of orthogonal invariance do not exist at all; I can even call this fact a characteristic of the general theory of relativity.[4]

He then concludes his first paragraph with the sentence, "It would be good to produce the mathematical proof of my conjecture."[5]

In answer to Hilbert, Klein adds the third and last part of his article, concluding with the words, "I would be very interested in seeing the development of the mathematical proof, which you announce at the end of the first paragraph of your answer."[6] When this article was reprinted in Klein's *Gesammelte mathematische Abhandlungen*, he added the following commentary: "The preceding development has since been provided by Miss Emmy Noether; see her paper on 'Invariant Variational Problems' in the *Göttinger Nachrichten* of 26 July 1918. I will return to this subject at the end of Section XXXII."[7] This is a clear statement that Noether found the key and mathematical proof for what Hilbert had only surmised.

In his summary of Klein's article for the *Jahrbuch über die Fortschritte der Mathematik*, which had no reason to be anything more than purely mathematical, the Prague mathematician Philipp Frank[8] emphasized this historical element in the last sentence of his review, "Both [Klein and Hilbert] refer explicitly to the collaboration of E. Noether in their research,"[9] an additional testimony to the fact that Noether's most distinguished colleagues recognized the importance of her contribution.

In 1924, six years after the publication of her article, Hilbert was even less disposed than Klein to give credit where it was clearly due, when he fused his articles of 1915 and 1917, with modifications, into a single article which he published in the *Mathematische Annalen*.[10] In it he cites Klein and Einstein quite often, as one would

---

[4] "Freilich behaupte ich dann, daß für die *allgemeine* Relativität, d. h. im Falle der *allgemeinen* Invarianz der Hamiltonschen Funktion, Energiegleichungen, die in Ihrem Sinne den Energiegleichungen der orthogonalinvarianten Theorien entsprechen, überhaupt nicht existieren; ja ich möchte diesen Umstand sogar als ein charakteristisches Merkmal der allgemeinen Relativitätstheorie bezeichnen," *ibid.*

[5] "Für meine Behauptung wäre der mathematische Beweis erbringbar," *ibid.*

[6] "Es würde mich aber sehr interessieren, die Ausführung des mathematischen Beweises zu sehen, den Sie am Ende des ersten Absatzes Ihrer Antwort in Aussicht stellen." Klein [1918a], p. 482, and *Gesammelte mathematische Abhandlungen*, vol. 1, p. 565. This sentence is quoted in translation in Rowe [1999], p. 215.

[7] "Besagte Ausführung ist inzwischen von Frl. E. Nöther geliefert worden, siehe deren Note über 'Invariante Variationsprobleme' in den Göttinger Nachrichten vom 26. Juli 1918. Ich komme hierauf am Schluß von XXXII zurück," *ibid.* Section XXXII is a reprint of Klein [1918b]. See *infra*.

[8] An Austrian contemporary of Noether, Frank (1884–1966) wrote his dissertation under Ludwig Boltzmann in Vienna and defended it in 1907. From 1912 to 1938 he was a professor of theoretical physics at the German University in Prague, where Einstein had recommended him to be his successor. He emigrated to the United States in 1938 and became a lecturer at Harvard.

[9] "Beide weisen ausdrücklich auf die Mitarbeiterschaft E. Noethers bei diesen Untersuchungen hin," *Jahrbuch über die Fortschritte der Mathematik*, vol. 46 (1916–1918), p. 1299.

[10] See Rowe [1999], pp. 227–228, who points out that this article contains "major alterations of the contents of the first note that no careful reader could possibly miss," and that it is therefore far from being, as Hilbert had claimed, a reprint of his papers of 1915 and 1917. In fact, Hilbert introduced his article [1924] as follows: "What follows is essentially a reprint of my two earlier communications including my remarks on them which F. Klein published in his communication

expect, but regarding the collaboration with Noether and her results, one finds only the following footnote:

> The proof of this theorem has been supplied by Emmy Noether in the general case (*Göttinger Nachrichten*, 1918, p. 235: "Invariant variational problems"). The identities indicated in Theorem 2 were already asserted in my first note, in reality only in the case where the invariant [i.e., the invariant Lagrangian] depends on the $g^{\mu\nu}$ and their [first] derivatives. But the procedure of the proof which is included and reproduced in the text is also valid for our general invariant, J. In their general form, the given identities were first deduced by F. Klein using the method of infinitesimal transformations.[11]

We now turn to Klein's articles and his later notes on them. In the introduction to his first communication, "On the differential laws for the conservation of momentum and energy in Einstein's theory of gravitation" [1918b], where he discusses the conservation laws in general relativity in the light of the work of Einstein and Hilbert, Klein declares, "As it will be evident from the presentation in what follows, I no longer need to perform any calculations, but need only use the most elementary formulas of the classical calculus of variations."[12] And in the last section he adds:

> I would be remiss if I did not thank Miss Noether once again for her active participation in my new work. She has by herself completely set out in proper form the mathematical ideas that I use in connection with the physics problems related to the integral I_1, [results] which will be presented in a note to appear shortly in these *Nachrichten*.[13]

The article ends with the following note: "I have already lectured on the principal theorems of Miss Noether on 23 July [1918] at the [Göttingen] Scientific Society." In the re-issue of the article in his *Gesammelte mathematische Abhandlungen*, Klein corrected the date from 23 to 26 July because the 23rd was actually the day when

---

[Klein 1918a]," ("Das Nachfolgende ist im wesentlichen ein Abdruck der beiden älteren Mitteilungen von mir über die 'Grundlagen der Physik' und meiner Bemerkungen dazu, die F. Klein in seiner Mitteilung 'Zu Hilberts erster Note über die Grundlagen der Physik' veröffentlicht hat [...]," Hilbert [1924], p. 1, and *Gesammelte Abhandlungen*, vol. 3, p. 258).

[11] "Den Beweis dieses Satzes hat allgemein Emmy Noether geliefert (Gött. Nachr. 1918 S. 235: ,Invariante Variationsprobleme'). Die in Theorem 2 angegebenen Identitäten sind in meiner ersten Mitteilung zwar nur für den Fall behauptet worden, daß die Invariante von den $g^{\mu\nu}$ und deren Ableitungen abhängt; aber das dort eingeschlagene und im Text reproduzierte Beweisverfahren gilt ebenso auch für unsere allgemeine Invariante J. In der allgemeinen Form sind die angegebenen Identitäten zuerst von F. Klein auf Grund der Methode der infinitesimalen Transformation abgeleitet worden." This footnote, Hilbert [1924], p. 6, and *Gesammelte Abhandlungen*, vol. 3 (1935), pp. 262–263, ends with a reference to Klein [1918a].

[12] "Ich habe, wie man sehen wird, bei der im folgenden zu gebenden Darstellung eigentlich überhaupt nicht mehr zu rechnen, sondern nur von den elementarsten Formeln der klassischen Variationsrechnung sinngemäßen Gebrauch zu machen," Klein [1918b], p. 172, and *Gesammelte mathematische Abhandlungen*, vol. 1, p. 568. This passage is translated in Pais [1982], p. 274.

[13] "Ich darf auch nicht unterlassen, für fördernde Teilnahme an meinen neuen Arbeiten wieder Frl. Nöther zu danken, welche die mathematischen Gedanken, die ich in Anpassung an die physikalische Fragestellung für das Integral I_1 benutze, ihrerseits allgemein herausgearbeitet hat und in einer demnächst in diesen Nachrichten zu veröffentlichenden Note darstellen wird," Klein [1918b], p. 189, and *Gesammelte mathematische Abhandlungen*, vol. 1, p. 584. This latter passage is translated in Mehra [1974], note 229, and in Byers [1996], a lecture delivered by the physicist Nina Byers (1930–2014) at a conference on the history of particle physics.

Noether presented her work to the Göttingen Mathematical Society, while the 26th was the day when he presented her work to the Scientific Society, and he added that Noether's article had since been published.[14] He then adds a 17-line commentary in which he gives a summary of Noether's results and emphasizes their significance:

> The principal theorem stated in Section 2 above is a special case of the following important theorem proven by Miss Noether (*loc. cit.*):
> "If an integral I is invariant under a group G_ρ (that is to say a continuous group with ρ essential parameters), then ρ independent linear combinations of the Lagrangian expressions become divergences."
> But regarding what in particular concerns Hilbert's assertion contained in Section XXXI (see pp. 561 and 565 of the present edition),[15] in the precise formulation of Miss Noether it becomes,
> "If an integral I is invariant under the group of translations, and if I is invariant with respect to an infinite group that contains the group of translations as a subgroup, then and only then do the energy relations become improper."
> Furthermore, Hilbert's proposition in Section XXXI, according to which there exist four relations among the field equations of [general] relativity theory, is also generalized by Miss Noether. Her theorem may be stated as follows:
> "If the integral I is invariant under a group that depends on ρ arbitrary functions and their derivatives up to order σ, then there exist ρ relations that are identically satisfied among the Lagrangian expressions and their derivatives up to order σ."[16]

In his second communication, "On the integral form of the conservation laws and the theory of the spatially closed universe" [1918c], Klein refers to his own 1910 article:

> In particular, I observed explicitly in 1910, in my description of the geometric foundations of the Lorentz group, that one should never speak about the theory of relativity[17] without

---

[14] "Die Hauptsätze von Frl. Nöther habe ich am 26. Juli der Gesellschaft der Wissenschaften vorgelegt. Die Note selbst ist weiterhin in den Göttinger Nachrichten 1918, S. 235–257, unter dem Titel ‚Invariante Variationsprobleme' erschienen," Klein, *Gesammelte mathematische Abhandlungen*, vol. 1, p. 585.

[15] Section XXXI is Klein [1918a], and Hilbert's assertion is the one quoted *supra*, p. 66.

[16] "Der vorstehend in §2 aufgestellte ‚Hauptsatz, ist ein besonderer Fall des folgenden von Frl. Nöther am angegebenen Orte bewiesenen weitreichenden Theorems:
‚Ist ein Integral I invariant gegenüber einer G_ρ (d. h. einer kontinuierlichen Gruppe mit ρ wesentlichen Parametern), so werden ρ linear unabhängige Verbindungen der Lagranschen Ausdrücke zu Divergenzen[.]'
Was aber inbesondere die in XXXI enthaltene Behauptung von Hilbert angeht (siehe S. 561 und 565 der vorliegenden Ausgabe), so ergibt sich als deren exakte Formulierung nach Frl. Nöther die folgende:
‚Gestattet ein Integral I die Verschiebungsgruppe, so werden die Energierelationen dann und nur dann uneigentliche, wenn I invariant ist gegenüber einer unendlichen Gruppe, die die Verschiebungsgruppe als Untergruppe enthält.'
Übrigens findet auch der Satz von Hilbert bzw. von XXXI, daß zwischen den feldgleichungen der Relativitätstheorie vier Relationen bestehen, bei Frl. Nöther seine Verallgemeinerung. Ihr theorem lautet so: ‚Ist das Integral I invariant gegenüber einer Gruppe mit ρ willkürlichen Funktionen, in der diese Funktionen bis zur σ-ten Ableitung auftreten, so bestehen ρ identische Relationen zwischen den Lagrangeschen Ausdrücken und ihren Ableitungen bis zur σ-ten Ordnung,'" Klein *Gesammelte mathematische Abhandlungen*, vol. 1, pp. 584–585. This commentary was translated by Basil Gordon in Byers [1999].

[17] I.e., the special theory of relativity developed in 1905.

further precision, but that one should always speak about the theory of invariants relative to a group. There are as many versions of the theory of relativity as there are groups.[18]

In fact, we see proof that Klein associated Noether with the generalization of general relativity theory that he outlines in this communication since, at the end of this paragraph, he adds the footnote,

> Also compare the communication by Miss Noether on the *Invariante Variationsprobleme* of 1918 in the *Göttinger Nachrichten* (the final footnote to this article).[19]

In the communication which Klein had presented to the Göttingen Mathematical Society on 10 May 1910, after general considerations on the relations between physicists and mathematicians, and the statement that, in principle, it is indispensable for mutual comprehension that the concepts of one discipline be translated into the language of the other, he had written,

> What modern physicists call the *theory of relativity* is the theory of invariants of the four-dimensional space-time domain, x, y, z, t, (of the Minkowski "universe") with respect to a particular group of collineations, more precisely, to the "Lorentz group";—or to a more general group, and, on the other hand:
>   One could indeed, if one really wanted to, replace the term "theory of invariants relative to a group of transformations" with the term "theory of relativity with respect to a group."[20]

This is the passage whose relevance Noether would stress eight years later in the last footnote of her paper,[21] in which she somewhat shortened Klein's sentence, thus rendering his idea even more striking.

In a letter to Einstein of 10 November 1918, Klein acknowledges Noether's help explicitly: "Meanwhile with Miss Noether's help, I have understood that the proof of the vectorial character of ε^σ from higher principles that I had sought was already

---

[18] "Insbesondere habe ich 1910 in meinem Vortrag über die geometrischen Grundlagen der Lorentzgruppe ausdrücklich bemerkt, daß man nie von Relativitätstheorie schlechtweg reden sollte, sondern immer nur von der Invariantentheorie relativ zu einer Gruppe.—Es gibt so viele Arten Relativitätstheorie als es Gruppen gibt," Klein [1918c], p. 399, and *Gesammelte mathematische Abhandlungen*, vol. 1, p. 590. Only four notes were added to this article for its re-issue in the *Gesammelte mathematische Abhandlungen*, and they are all cross-references to material in that edition.

[19] "Vergleiche auch die Mitteilung über ,invariante Variationsprobleme' von Frl. Noether im Jahrgang 1918 dieser Nachrichten (Schlußbemerkung daselbst)," Klein [1918c], p. 399, and *Gesammelte mathematische Abhandlungen*, vol. 1, p. 590.

[20] "Was die modernen Physiker *Relativitätstheorie* nennen, ist die Invariantentheorie des vierdimensionalen Raum-Zeit-Gebietes, x, y, z, t (der Minkowskischen ,Welt') gegenüber einer bestimmten Gruppe von Kollineationen, eben der ,Lorentzgruppe';—oder allgemeiner, und nach der anderen Seite gewandt:
Man könnte, wenn man Wert darauf legen will, den Namen ,Invariantentheorie relativ zu einer Gruppe von Transformationen'sehr wohl durch das Wort ,Relativitätstheorie bezüglich einer Gruppe'ersetzen," Klein [1910], p. 287, and *Gesammelte mathematische Abhandlungen*, vol. 1, p. 539.

[21] That is, note 1 on p. 257 of Noether [1918c] (note 27, p. 22 of the above translation). See Chap. 2, p. 64.

given by Hilbert in pp. 6, 7 of his first note, in any case in a version that does not bring out the essential point."[22]

One would expect to find a reference to Noether's article in the second volume of the "Lectures on the Development of Mathematics in the Nineteenth Century" [1927] that Klein wrote shortly before his death, 22 June 1925, and which was prepared for publication by Richard Courant. In fact, while "Noether"—the mathematician Max Noether, Emmy Noether's father—is mentioned fifteen times in volume 1, she is mentioned only twice, in the second volume, the first time (p. 186) without any reference, for her contribution to the theory of differential invariants in geometry and to the study of surfaces of constant curvature, and the second time (p. 199), for her contribution to the study of differential invariants of quadratic differential forms, with references to her [1918b], to an article by Hermann Vermeil [1919],[23] and to Supplement 1 to the fifth edition of Weyl's *Raum, Zeit, Materie* (1925) [*sic* for 1923]. According to Kastrup [1987], p. 124, Vermeil's article consisted of "details [...] [that] were worked out according to Noether's ideas" contained in her [1918b]. So Klein could not avoid citing Noether here. However, when he discusses the conservation laws of mechanics on pp. 56–59, he refers to Jacobi and cites Herglotz ([1911], pp. 512–513), Engel [1916], Schütz [1897] and, for generalizations, Heinrich Burckhardt,[24] Eduard Study[25] and Weitzenböck, without mentioning Noether. His omission of Noether's contribution to this field is unexpected.

Einstein — There are two passages from Einstein's correspondence that must be quoted because they concern Noether. In a letter of 24 May 1918 Einstein writes to Hilbert,

> Yesterday I received from Miss Noether a very interesting paper on the generation of invariants.[26] I am impressed by the fact that these things can be understood from so general

---

[22] "Inzwischen habe ich mit Hülfe von Frl. Noether verstanden, daß der Beweis für den Vektorcharakter von ε^σ aus ,höheren Prinzipien', wie ich ihn suchte, schon von Hilbert auf p. 6, 7 seiner ersten Note gegeben worden ist, allerdings in einer Redaktion, die das Wesentliche nicht hervorkehrt," *Collected Papers* 8B, no. 650, p. 942; 8 (English), p. 692. This letter was mentioned, *supra*, Chap. 1, p. 43.

[23] Vermeil (1889–1959) was Klein's assistant between 1919 and 1921, in fact his last assistant, and, with Robert Fricke (1861–1930), was responsible for the publication of the second volume of Klein's *Gesammelte mathematische Abhandlungen* in 1922. Erich Bessel-Hagen (1898–1946) continued Vermeil's work as editor after 1921, and the third and final volume of Klein's works appeared in 1923, edited by Fricke, Vermeil and Bessel-Hagen. In a letter to Einstein of 5 November 1918 (*Collected Papers* 8B, no. 645, pp. 936–937; 8 (English), pp. 687–688), Klein mentions the assistance that Vermeil supplied him regarding the question of Hilbert's energy vector.

[24] Burckhardt (1861–1914) had defended his thesis on invariants and algebraic integrals in Munich in 1897.

[25] Study (1862–1930) was a professor at Bonn from 1903 to 1927. His work deals not only with invariant theory but also with projective geometry, transformation groups and mechanics. He had been a Privatdocent at Leipzig when Lie came as a professor in 1886, and "became interested in the connection between transformation groups and the theory of invariants." (Letter from Lie to Klein, ca. 1888, from the Göttingen archives, quoted in Hawkins [2000], p. 236.) See Fisher [1966] and Hawkins [1998].

[26] Einstein is referring to her [1918b].

a point of view. It would have done the Old Guard of Göttingen no harm to be sent back to school under Miss Noether. She really seems to know her trade![27]

Einstein would be still more explicit on receiving Noether's new article, the one that is being studied here. In a letter to Klein of 27 December 1918, which is preserved in the Göttingen archives, after congratulating Klein on his "fine" recent jubilee, and thanking him for having sent him his elegant proof of the vectorial character of the quantities introduced by Hilbert, he writes,

> What brings me to write to you today is, however, another matter. After having received Miss Noether's new article,[28] I once again feel that refusing her the right to teach[29] is a great injustice. I would be very favorable to taking energetic steps [on her behalf] before the ministry. If you do not think that this is possible, then I will go to the trouble of doing it alone.[30]

The affair seemed so important to him that he added, "Unfortunately, I must leave on a trip for a month. But I urge you to send me a brief account [of what you will have done] on my return. If something must be done before that, please feel free to use my signature."[31]

Weyl — In his *Raum, Zeit, Materie* [1918b], more precisely, in the subsequent, revised editions[32] in which he obtained the laws of conservation of energy and momentum from a calculation of infinitesimal vertical variations, Weyl mentioned Noether only in a note. He first writes:

---

[27] "Gestern erhielt ich von Fr. Nöther eine sehr interessante Arbeit über Invariantenbildung. Es imponiert mir, dass man diese Dinge von so allgemeinem Standpunkt übersehen kann. Es hätte den Göttinger Feldgrauen nichts geschadet, wenn sie zu Frl. Nöther in die Schule geschickt worden wären. Sie scheint ihr Handwerk gut zu verstehen!" Einstein, *Collected Papers* 8B, no. 548, pp. 774–775; 8 (English), pp. 568–569. In 1972, this letter was in the possession of Helen Dukas. (See, *infra*, note 55.) This text is quoted in Kimberling [1981], p. 13 and, in translation, in note 5 of p. 46, and also by Rowe [1999], p. 213. According to Freeman J. Dyson, in a letter to Clark H. Kimberling quoted in the addendum to Kimberling's article [1972], *Feldgrauen* was slang for "warriors".

[28] It is clear from the date of this letter that the "new article" is Noether [1918c].

[29] Einstein uses the technical term, *venia legendi*.

[30] "Was mich heute zum Schreiben veranlasst, ist etwas anderes. Beim Empfang der neuen Arbeit von Frl. Noether empfinde ich es wieder als grosse Ungerechtigkeit, dass man ihr die venia legendi vorenthält. Ich wäre sehr dafür, dass wir beim Ministerium einen energischen Schritt unternähmen. Halten Sie dies aber nicht für möglich, so werde ich mir allein Mühe geben," Einstein, *Collected Papers* 8B, no. 677, pp. 975–976; 8 (English), p. 714. A translation of this passage appears in Rowe [1999], p. 198.

[31] "Leider muss ich für einen Monat verreisen. Ich bitte Sie aber sehr, mit kurz Nachricht zu geben bis zu meiner Rückkehr. Wenn vorher etwas gemacht werden sollte, so bitte ich Sie, über meine Unterschrift zu verfügen," *ibid.*

[32] Weyl had written to Einstein on 1 March 1918, *Collected Papers* 8B, no. 472, pp. 663–664; 8 (English), p. 487, that he was going to send him the proofs of his book, and on reading them, Einstein's reaction on March 8 was enthusiastic (letter to Weyl, *Collected Papers* 8B, no. 476, pp. 669–670; 8 (English), p. 491). On 16 November the press-run was nearly sold out, so Weyl authorized a reprint of 600 copies (2nd, unchanged edition, 1919) while preparing the considerably revised 1919 edition (3rd edition). See Weyl's letter to Einstein, *Collected Papers* 8B, no. 657, pp. 948–950; 8 (English), p. 696.

We shall use a beautiful idea whose origin may be found in the masterful work of Lagrange and whose most perfect form has been developed by Klein.[33]

He then directs the reader to his note 5 to chapter IV printed at the end of the book where he supplies a reference to Klein [1918b] followed by an additional reference, apparently slightly less important in his view:

Also see the general calculations of E. Noether, *Invariante Variationsprobleme*, in the same journal.[34]

In fact, the article by Klein that Weyl cites in this note is the one which we cited above (Klein [1918b], see, *supra*, p. 68) which contains Klein's thanks to Noether for "her active participation" and for having "set out in proper form the mathematical ideas that [he] use[d] [in the article]." The brevity with which Weyl wrote about Noether in his *Raum, Zeit, Materie* is very surprising.

Further on, Weyl evokes "the question whether proper theorems of conservation may actually be set up" in general relativity, and indicates that he will postpone this discussion to a later section of his book.[35] There he discusses the meaning of the "components of the energy-density of the gravitational field,"[36] and in his note 27 to chapter IV printed at the end of the book he supplies references to Einstein [1916a] and [1918], and once again to Klein, this time [1918c], while neglecting to cite Noether.[37]

On 24 March 1918, Einstein wrote to Klein, "Recently I have received the proofs of a book by H. Weyl on the theory of relativity which has made a deep impression

---

[33] "Sondern bedienen uns dazu der folgenden schönen Überlegung, deren Keime bei Lagrange zu finden sind, die aber in volkommenster Form von F. Klein auseinandergesetzt wurde," Weyl [1918b], 3rd ed., 1919, p. 199, 4th ed., 1921, p. 211. This passage is not in the earlier editions. The English translation, "Space, Time, Matter" (1922, p. 233), has "But we must apply the following elegant considerations, the nucleus of which is to be found in Lagrange, but which were discussed with due regard to formal perfection by F. Klein." The French translation, "Temps, Espace, Matière" (1922 and 1958 reprint, p. 204) has "Nous nous servirons d'une idée fort belle dont les germes se trouvent dans l'œuvre magistrale de Lagrange et dont la forme la plus parfaite a été développée par Klein." Weyl did not approve the French translation of his book and he expressed his disapproval at the end of the preface of the fifth German edition (1923), "it is so 'free' in places that I refuse to take reponsibility for its contents." ("[Sie] ist allerdings stellenweise so ,frei', daß ich mich genötigt sehe, für ihren Inhalt jede Verantwortung abzulehnen.") See Coleman [1997], p. 10.

[34] "Vgl. dazu die allgemeinen Formulierungen von E. Noether, *Invariante Variationsprobleme*, am gleichen Ort," Weyl [1918b], 3rd ed., 1919, p. 266, 4th ed., 1921, p. 292. This note does not appear in the earlier editions. The English translation (1922, p. 322), has "Cf., in the same periodical, the general formulations given by E. Noether, *Invariante Variationsprobleme*." The French translation (1922 and 1958 reprint, p. 204) has "Voir aussi à ce propos les calculs très généraux de E. Noether: *Invariante Variationsprobleme*, id."

[35] " Die Frage, ob sich wirkliche Erhaltungssätze aufstellen lassen, wird erst in § 32 geprüft werden." Weyl [1918b], 3rd ed., 1919, p. 202; English translation, 1922, p. 236; French translation, 1922, p. 285. The section where Weyl studies the question of conservation laws in general relativity is 32 in the 3rd edition, 33 in the 4th, and 37 in the 5th.

[36] "Die Komponenten der Energiedichte des Gravitationsfeldes," Weyl [1918b], 3rd ed., 1919, p. 233, 4th ed., 1921, p. 246; English translation, 1922, p. 271.

[37] Weyl [1918b], 3rd ed., 1919, p. 267, 4th ed., p. 293; English translation, 1922, p. 323.

on me. It is admirable how he masters the subject. He obtains the law of energy of matter by the same variational technique that you use in your recently published note."[38] But three months later, an extremely favorable review of the first edition of *Raum, Zeit, Materie*[39] contains one of the few reservations that Einstein expressed in its regard, "For the sake of completeness, I must say that I am not entirely in agreement with the author regarding his interpretation of the energy theorem, as well as the relation that exists between the predictions of theoretical physics and reality."[40] It is clear that Einstein's reservation had to do with the difficult problem of the conservation of energy on which Noether's second theorem would soon shed some light.

In an article which is considered to be the precursor of gauge theories [1918a], Weyl showed that the conservation of charge in electromagnetism theory was in fact a consequence of the invariance of the equations of electromagnetism under the action of the group of scaling transformations, which depend on "an arbitrary positive function of the positions" ("eine beliebige positive Ortsfunktion"), which invariance he first called an "invariance of calibration" or "scale invariance" (*Maßstab-Invarianz*),[41] and subsequently, a "gauge invariance" (*Eich-Invarianz*).[42] This fact can be seen as an application of Noether's second theorem, where gauge invariance of the second type (see, *infra*, Chap. 6, p. 123) implies the existence of an improper conservation law, but he made no allusion to that. He wrote:

> We shall show that *just as*, according to the researches of Hilbert [1915], Lorentz [1915] [1916], Einstein [1916b], Klein [1918a] and the author,[43] *the four conservation laws of*

---

[38] "Ich habe in letzter Zeit Korrekturen eines Buches von H. Weyl über Relativitätstheorie gelesen, die grossen Eindrück auf mich gemacht haben. Es ist bewundernswert wie er den Stoff meistert. Den Energiesatz der Materie leitet er mit demselben Variations-Kunstgriff ab wie Sie in Ihrer neulich erschienenen Note," letter cited *supra*, Chap. 1, note 57.

[39] This review, published 21 June 1918 in *Die Naturwissenschaften*, 6 (1918), p. 373, is printed in his *Collected Papers* 7, no. 10, pp. 78–80; 7 (English), pp. 62–63. Einstein writes that "every page shows the amazingly steady hand of the master who has penetrated the subject matter from the most diverse angles" ("denn jede Seite zeigt die unerhört sichere Hand des Meisters, der den Gegenstand von der verschiedensten Seiten durchdrungen hat"), p. 79, English, p. 62.

[40] "Der Vollständigkeit halber sei erwähnt, daß ich mit dem Verfasser nicht ganz übereinstimme bezüglich der Auffassung des Energiesatzes sowie des Verhältnisses, welche zwischen den Aussagen der theoretischen Physik und der Wirklichkeit besteht," *ibid.*, p. 78, English, p. 63.

[41] Weyl [1918a], p. 475, and *Gesammelte Abhandlungen*, vol. 2, p. 38. The term "Maßstab-Invarianz" also appears in Weyl [1918c], p. 404, and *Gesammelte Abhandlungen*, vol. 2, p. 15.

[42] For the expression "Eichinvarianz" or its hyphenated form, see Einstein to Weyl, 29 November 1918, *Collected Papers* 8B, no. 661, p. 954; 8 (English), p. 700 (with the spelling "Aich-Invarianz," translated as "gauge invariance") and Weyl [1919], p. 114, and *Gesammelte Abhandlungen*, vol. 2, p. 75; also see p. 59, where he writes *Eichung*. I thank Prof. Erhard Scholz for these references. In the letter to Besso written from Zurich at the end of 1913 or in early January 1914 (Einstein and Besso [1972], no. 9, p. 50; *Collected Papers* 5, no. 499, p. 588; 5 (English), p. 373) Einstein alluded to a scalar "accentuating factor" (",Betonungsfaktor' ϕ (skalar)") for the energy tensor, but it is unrelated to the later concept of gauge factor. Kastrup wrote ([1987], p. 124) that Bessel-Hagen was the first to have recognized the role of gauge invariance in the law of charge conservation in electrodynamics, a fact which had already been observed by Weyl.

[43] Weyl [1917], pp. 121–125.

matter (of the energy-momentum tensor) *are connected with the invariance of the action with respect to coordinate transformations*, expressed through four independent functions, the conservation law of electricity is connected with the new scale invariance, expressed through a fifth arbitrary function.[44]

Considering the theory of Weyl [1919], Pauli mentioned, concerning the law of charge conservation, that it is "formally exactly on the same footing as the law of conservation of energy,"[45] but he concluded that Weyl's theory failed to resolve the problem of the structure of matter.[46] In fact, Weyl's theory of matter in 1918 and 1919 was only a preliminary to his *Elektron und Gravitation* [1929] which, by passing to quantum theory, became the first of the numerous, later theories based on the principle of gauge invariance.[47] In that famous article, Weyl obtained "the quasi-conservation law of energy and momentum"[48] in general relativity from the invariance properties of the action integral before showing how the gauge invariance of the Lagrangian of electromagnetism implied the conservation of charge. He referred to the fifth edition of his *Raum, Zeit, Materie* three times, but not once to Noether, even though his proofs are very close to hers. In fact, in his analysis of the prehistory and the history of gauge theories, Lochlainn O'Raifeartaigh [1997], p. 116, observes explicitly, "Thus, although Weyl did not refer to Noether, this particular result of his was actually a special case of Noether's theorem."

Many years later, stimulated by an article by Constantin Carathéodory[49] [1929], Weyl published another article [1935b] in which he first gave a short account of the general theory of the calculus of variations. While in section 7 he dealt with "invariance," the word there designates the tensorial nature of the Lagrangian—it is in fact a tensor density—and of other quantities that appear in the calculus of variations,[50] and he did not allude to Noether's work here either.

---

[44] "Wir werden nämlich zeigen: *in der gleichen Weise*, wie nach Unteruschungen von Hilbert, Lorentz , Einstein, Klein und dem Verfasser *die vier Erhaltungssätze der Materie* (des Energie-Impuls-Tensors) *mit der*, vier willkürliche Funktionen enthaltenden *Invarianz der Wirkungsgrösse gegen Koordinatentransformationen zusammenhängen, ist mir der hier neu hinzutretenden*, eine fünfte willkürliche Funktion hereinbringenden "Maßstab-Invarianz" [...] das Gesetz von der Erhaltung der Elektrizität verbunden," Weyl [1918a], p. 473, reprinted in Lorentz *et al.* [1922], 5th ed., 1923, p. 156, and in *Gesammelte Abhandlungen*, vol. 2, p. 37. We have adapted the translation of O'Raifeartaigh [1999], p. 32. See also Lorentz *et al.* [1923], p. 212.

[45] Pauli [1921], p. 768, [1958], p. 200.

[46] *Ibid.*, p. 770, [1958], p. 202.

[47] For the evolution of Weyl's thought, see, in his *Gesammelte Abhandlungen*, vol. 2, p. 42, his commentary on his own 1918 article answering Einstein's objections. See O'Raifeartaigh [1997], chapter 4, for the role of the work of Erwin Schrödinger, Fritz London, Paul A. M. Dirac and Vladimir Fock in the development between 1918 and 1929 of the gauge theory of electromagnetism.

[48] "Der Quasi-Erhaltungssatz von Energie und Impuls," Weyl [1929], p. 343, and *Gesammelte Abhandlungen*, p. 257.

[49] See, *infra*, p. 99.

[50] Jets of maps—which would be introduced by Charles Ehresmann (see, *infra*, Chap. 5, p. 110)— appear already in Weyl [1935b] without a name.

3.2 The Eulogies of 1935

After Noether's death, 14 April 1935, several eulogies were published. In addition to those we shall analyze below, those by José Barinaga, A. Sagastume Berra and V. Kořínek should be mentioned.[51] Barinaga, in his brief tribute [1935], explains that Noether "subsequently extended her ideas to the invariants of differential expressions, and in particular to those of the calculus of variations (1918)."[52]

Einstein — In a letter to the editor of the *New York Times* (4 May 1935) that appeared under the title, surely supplied at least in part by the editors, "The Late Emmy Noether, Professor Einstein Writes in Appreciation of a Fellow-Mathematician," Einstein wrote, "In the judgment of the most competent living mathematicians, Fräulein Noether was the most significant creative mathematical genius thus far produced since the higher education of women began."[53] This is a vivid tribute to the "enormous importance" of her work "in the realm of algebra," but does not enter into detail or allude to the 1918 article even though it had made, as we have just shown, an important contribution to the early debates on general relativity. One could not expect more scientific detail in a *New York Times* article of that period, although the *Times* had already engaged in 1927 the highly competent Waldemar Kaempffert as its science correspondent. The tribute also contains an explicit criticism of the way Noether had been treated in Germany, a subject which is rather unexpected in the *New York Times* of the mid-1930s which was still prudish about Jewish matters.

Kimberling [1972] claims that a note found in the archives of the *Bryn Mawr Alumnæ Bulletin* asserts that Einstein never met Noether, and further that his tribute to her was inspired and possibly even drafted by Weyl. In fact, it is likely that Einstein met Noether in Göttingen since she was already in residence when he came to deliver lectures on his theory of gravitation at the end of June 1915.[54] It is probable that she attended at least one of these lectures and that they would have met on that occasion. While it is not unlikely that Weyl wrote the tribute to Noether for Einstein, it seems not to have been the case. In the addendum to [1972], Kimberling reports that Freeman Dyson denied that claim in a short letter to him, "Miss Dukas[55] has the original German draft of this letter [...] written by Einstein himself at the request

---

[51] These eulogies are identified in Poggendorff, *s. v.*, Noether, in Dick [1970] and [1981], p. 187, and in Kimberling [1981]. We have also seen in the Bryn Mawr archives a very touching, autograph tribute to Noether by her colleague, Margarita Lehr, that was apparently pronounced at a memorial meeting at the college, but does not discuss Noether's mathematics in any detail.

[52] "Después extiende sus ideas a los invariantes de expresiones diferenciales, y en particular a los del cálculo de variationes (1918)," Barinaga [1935], p. 62. José Barinaga Mata (1890–1965) was then chief editor of the *Revista Matematica Hispano-Americana*, and he became president of the Spanish Mathematical Society (*Sociedad Matemática Española*) two years later.

[53] The letter is quoted in Dick [1970], p. 37, and [1981], p. 92, in Kimberling [1972] and extracts are quoted in Pais [1982], p. 276, and Byers [1996], p. 956. See also Byers [2006].

[54] See, *supra*, Chap. 1, p. 40.

[55] Helen Dukas was Einstein's secretary at Princeton. In 1972, she was in charge of the Einstein archive at the Institute for Advanced Study, and she would, in 1979, together with Banesh Hoffmann, edit and translate a small part of Einstein's correspondence.

of Weyl." Dyson then quotes the passage of the letter of 24 May 1918 from Einstein to Hilbert that we have already cited,[56] and concludes,

> From the letter you can see that, while it may be true that Einstein and Emmy Noether never met (Miss Dukas is not sure about this), Einstein certainly knew her work well and understood its importance early and at first hand.

Another, contradictory testimony has been given by Ruth Stauffer McKee, who had been Noether's only doctoral student at Bryn Mawr. On 17 October 1972, after having read the article by Kimberling and its addendum in the *American Mathematical Monthly*, she wrote to the editor, Harley Flanders, with a copy to Kimberling, to deny that any such note could be found in the files of the *Alumnæ Bulletin*,[57] and she emphasized the friendly and exciting atmosphere at Princeton in the presence of Solomon Lefschetz, Weyl and John von Neumann, where Noether had attended the welcoming party for Einstein in December 1933 and where she had obviously met him often. McKee then gives her own version of the background of the letter to the *New York Times*, "it was [...] obvious to all the mathematicians that Weyl should write the obituary—which he did. He, furthermore, sent it to the New York Times, the New York Times asked who is Weyl? Have Einstein write something. He is the mathematician recognized by the world. This is how Einstein's article appeared. It was most certainly 'inspired' by Weyl's draft." Whether or not it was 'inspired' by Weyl's text, Einstein certainly reiterated the appreciation of Noether that we have just cited from his correspondence.[58]

Other tributes to Noether by mathematicians appeared in volumes dedicated to her memory, but we shall discuss only those that deal with the aspect of her work which interests us here.

Weyl — Weyl pronounced two eulogies at Bryn Mawr, one very personal in German, at the burial,[59] and another, in English, equally touching but more scientific, at a memorial ceremony at Bryn Mawr College on 26 April 1935. In his tribute at the Bryn Mawr ceremony, he said that, in 1916,

> Hilbert at that time was over head and ears in the general theory of relativity, and for Klein, too, the theory of relativity and its connection with his old ideas of the Erlangen program brought the last flareup of his mathematical interests and mathematical production. The second volume of his history of mathematics in the nineteenth century [Klein [1927]] bears witness thereof. To both Hilbert and Klein Emmy was welcome [in Göttingen] as she was able to help them with her invariant theoretic knowledge.[60]

---

[56] *Supra*, p. 71 and note 27.

[57] "A careful check in the files by the staff of the [Bryn Mawr Alumnæ] Bulletin assured me that there was nothing in the files of the Bulletin to even imply that 'Mr. Einstein had never met Miss Noether.'" I thank Prof. Peter Roquette for communicating to me the contents of this letter, a copy of which he had received from the personal archive of Prof. Kimberling, and which he has now published [2008].

[58] The authorship of this letter was discussed by Roquette [2008].

[59] Roquette discovered this eulogy only recently and published it in an English translation [2008].

[60] Weyl [1935a], p. 206, and *Gesammelte Abhandlungen*, vol. 3, p. 430. See Chap. 1, p. 44, note 71.

He continued:

> For two of the most significant sides of the general relativity theory she gave at that time the genuine and universal mathematical formulation.

There follows a very short analysis of Noether's [1918b] article, "First the reduction of the problem of differential invariants to a purely algebraic one by use of 'normal coordinates,'" and then a summary of the second theorem of [1918c]:

> Second, the identities between the left sides of Euler's equations of a problem of variation which occur when the (multiple) integral is invariant with respect to a group of transformations involving arbitrary functions (identities that contain the conservation theorem of energy and momentum in the case of invariance with respect to arbitrary transformations of the four world coordinates).

Thus Weyl, who respected Noether to the point of admitting in this eulogy that he "was ashamed to occupy such a preferred position [in Göttingen] beside her whom [he] knew to be [his] superior as a mathematician in many respects,"[61] only mentioned ever so briefly the aspect of Noether's work which would have such a profound influence on the development of mathematical physics.

Van der Waerden — A eulogy written by the Dutch mathematician Bartel van der Waerden (1903–1996) was published in the *Mathematische Annalen* in 1935, i.e., after the Nazi party had seized power in Germany, even though Noether was Jewish. At that time, van der Waerden had been on the main editorial board of the *Mathematische Annalen* for a year and, although until April 1940 there was still no law forbidding the publication of the works of Jewish authors, it was then already daring to publish an article about a Jewish mathematician. It seems that the importance of Noether's mathematical legacy was such that a suitable tribute had to be published, even in Nazi Germany.[62] In this detailed eulogy, which is both personal and scientific, van der Waerden speaks briefly about Noether's early work: "Under the influence of Klein and Hilbert, who were both at the time much involved in the general theory of relativity, she completed her articles concerning differential invariants which have acquired great importance in this field,"[63] and he mentions the *Invariante Variationsprobleme* in a single sentence, "[...] In the second [article], she used

---

[61] Weyl [1935a], p. 208, and *Gesammelte Abhandlungen*, vol. 3, p. 432. Weyl had been named to succeed Hilbert at Göttingen in 1930.

[62] The reorganization of the editorial board followed a conflict between Hilbert and Brouwer in 1928. Its managing editor since 1905, Otto Blumenthal, subject to the anti-Jewish laws, offered his resignation in a letter to Hilbert dated 11 November 1933 (Bergmann *et al.* [2011]), but his name remained on the front page until 1939. He fled to Holland from where he was eventually deported to the Theresienstadt concentration camp, where he died. Segal observes [2003, p. 307] that publishing an obituary for Emmy Noether in 1935 in a German journal had been "a courageous act for even (or perhaps especially) a foreigner."

[63] "Unter dem Einfluß von Klein und Hilbert, die sich in dieser Zeit beide sehr mit der allgemeinen relativitätstheorie beschäftigten, kamen ihre Arbeiten über Differentialinvarianten [1918b] [1918c] zustande, welche für dieses Gebiet von großer Wichtigkeit geworden sind," van der Waerden [1935], p. 470. This text is quoted in Dick [1970], pp. 47–52, and translated in Dick [1981], pp. 100–111, and, by Christina M. Mynhardt, in Brewer and Smith [1981], pp. 93–98.

the methods of the formal calculus of variations for the formation of differential invariants."[64]

Alexandrov — Pavel Sergeievich Alexandrov was personally acquainted with Noether, having seen her almost daily during his visits to Göttingen in 1923 and 1924 as well as during the summers of 1926 and 1927, when he and Heinz Hopf conducted a seminar which Noether attended. He visited Noether every summer in Göttingen until 1932, and he received her in Moscow during the winter of 1928–1929. The very heartfelt tribute to her that he pronounced 5 September 1935 before the Moscow Mathematical Society, of which he was then president, was published in 1936. In it Alexandrov describes her work before 1920 in very few words:

> [...] Emmy Noether was fully capable of mastering such methods [of computations and of algorithms]. This is proved not only by her first dissertation, which in actual fact was not a major work, but also by her subsequent papers on differential invariants (1918) which have become classics. But in these papers we already see the fundamental characteristic of her mathematical talent, the striving for general formulations of mathematical problems and the ability to find the formulation which reveals the essential logical nature of the question, stripped of any incidental pecularities which complicate matters and obscure the fundamental point. [...] These results [concerning Hilbert's concrete algebraic problems] and her work on differential invariants would have been enough by themselves to earn her the reputation of a first class mathematician and are hardly less of a contribution to mathematics than the famous research of S. V. Kovalevskaya.[65]

Alexandrov then contrasts "these early works, important though they were in their concrete results," with "the main period in her research, beginning in about 1920," and he does not return to the subject of differential invariants in the remaining ten pages of his eulogy.

3.3 Personal Recollections

We now review reminiscences of Noether and appraisals of her 1918 paper by several of her younger contemporaries.

Wigner — Born in Budapest, Eugene Wigner (1902–1995) obtained a doctorate in engineering in Berlin in 1925. He spent the 1927–1928 academic year in Göttingen at the invitation of Hilbert who was ill when he arrived. This was the year that Noether

---

[64] "[...] In der zweiten werden die Methoden der formalen Variationsrechnung zur Bildung von Differentialinvarianten herangezogen," *ibid.*

[65] Alexandrov [1936]. We have used the translation that Neal and Ann Koblitz prepared for the *Gesammelte Abhandlungen / Collected Papers* of Noether [1983], pp. 1–11. This text also appears in English translation in Dick [1981], pp. 153–179, and has been translated by E. L. Lady, who used the spelling "Paul Alexandroff," for the volume edited by Brewer and Smith [1981], pp. 99–111. Long extracts have been published by Kimberling [1972]. Also see Alexandrov's "Pages from an autobiography" [1979] and, for the importance of the Noether–Alexandrov connection for the introduction of group theory into combinatorial topology, see Jacobson's preface to Noether's *Gesammelte Abhandlungen / Collected Papers*, pp. v–vi.

gave a course on the representations of algebras, whose content would constitute the article "Hypercomplex quantities and the theory of representations" [1929] that she wrote on the basis of notes that had been taken by van der Waerden.[66] While in Göttingen, Wigner worked mostly with the physicists, and even if he met Noether who lectured there and was, by Weyl's testimony, a strong personality who attracted an entourage of Göttingen students, he does not seem to have had any direct knowledge of her work on conservation laws, nor of her research of the 1920s on representation theory. It is remarkable that Wigner became aware of the relevance for his own research of either aspect of Noether's œuvre only decades later.

There is a symmetrical story from the other side of the mathematics/physics divide in the *Souvenirs d'apprentissage* [1991] of André Weil (1906–1998), who spent a part of 1926 in Göttingen. He writes about Noether's courses and the "conversations with members of her entourage" that introduced him "to what one began to call 'modern algebra,'" but he is surprised that, among the mathematicians, he did not detect "the least inkling" of the effervescence of the world of the physicists who were then and there "in the process of creating quantum mechanics."[67] Weil's eye-witness account suggests that, in contrast to the period in which Noether formulated and proved her theorems, when the mathematicians Klein and Hilbert, and the physicist Einstein corresponded while trying to understand the mathematical implications of general relativity, mathematicians and physicists in the late 1920s had few scientific exchanges, which may explain how, despite his scientific activity in Göttingen, Wigner could have remained ignorant of Noether's work of 1918.

While Noether and her predecessors used the Lagrangian formulation of classical and relativistic mechanics to obtain a correspondence between conservation laws and the invariance properties of the systems under consideration, Wigner, in his 1927 article in the *Göttinger Nachrichten*, obtained a similar correspondence for quantum mechanics using the linearity of Hilbert space, where all states are linear superpositions of pure states.[68] It is in that article that he introduced the conservation law which is associated with the principle of parity conservation, which principle has no analogue in classical (as opposed to quantum) physics. The parity symmetry that he treated in this paper is a discrete symmetry to which Noether's infinitesimal

---

[66] Regarding the importance of this article and Noether's contribution to the theory of algebras, see Curtis [1999].

[67] "As I learned much later, the physicists' world was then in great effervescence in Göttingen; they were in the process of creating quantum mechanics; it is quite remarkable that I did not have the least inkling of this. [...] Emmy Noether played the part of a protective mother hen [...] Had they been less disorganized, her lectures could have been useful, but it was there however, and in conversations with members of her entourage, that I started to learn what was beginning to be referred to as 'modern algebra' [...]" ("Comme je l'ai su bien plus tard, le monde des physiciens était alors en pleine effervescence à Göttingen ; ils étaient en train d'accoucher de la mécanique quantique ; il est assez remarquable que je n'en aie pas eu le moindre soupçon. [...] Emmy Noether jouait le role de mère poule [...] Moins désordonnés ses cours auraient pu être utiles, mais c'est là néammoins, et dans des conversations avec son entourage, que je m'initiai à ce qu'on commençait à appeler 'l'algèbre moderne' [...]," Weil [1991], pp. 51–52).

[68] See, e.g., Wigner [1967], pp. 10–11.

methods do not apply. Thus his result was an independent discovery that did not imply an acquaintance with Noether's theorems.

In his *Symmetries and Reflections* [1967] Wigner re-issued several earlier essays dealing with the relations between symmetries and conserved quantities in which he had emphasized above all the role of invariance principles in the formulation of the laws of nature, the fact that a law of physics can be valid only if it is compatible with the assumed invariance properties. In the first essay of that collection, "Invariance in Physical Theory" [1949], he wrote that, after Einstein's work on special relativity, "it is now natural for us to try to derive the laws of nature and to test their validity by means of the laws of invariance [. . .]. Once the fundamental equations are given, the principles of invariance furnish, in the form of conservation laws and otherwise, powerful assistance toward their solution."[69] He then states that, for classical mechanics, the results concerning the derivation of conservation laws from the invariance of the equations under infinitesimal translations and rotations are "due to F. Klein's school," and he cites F. Engel [1916], G. Hamel [1904a] and E. Bessel-Hagen [1921], omitting Noether although she clearly was the author of the essential step in this procedure.[70]

In his second essay, "Symmetry and Conservation Laws" [1964a], Wigner implied that, when the law of conservation of angular momentum became very important in Bohr's theory of the atom, physicists assumed the validity of that conservation law without knowing that Hamel had justified it as early as 1904 by the invariance of the system under rotations.[71] He also referred to Engel [1916], but omits both Bessel-Hagen and, again, Noether. He observes that the change in approach to those questions occurred between the beginning of the century and the period when he was writing, when "the relation between laws of conservation and invariance principles came to be accepted—almost too generally." To support this observation, he cited one of his earlier articles [1954] in which he had shown that the relation between symmetries and conservation laws could fail for physical systems whose equations could not be written in Hamiltonian form,[72] and he had argued that this is not in fact in contradiction with the results of Hamel, Engel and Bessel-Hagen.

In a third essay, "Events, Laws of Nature, and Invariance Principles" [1964b],[73] Wigner emphasized once more the importance of deriving conservation laws from the invariance properties of physical systems, classical and, above all, quantum, but this time he mentioned Noether [1918c] in the company of Hamel [1904a],

---

[69] This text was an address delivered at a ceremony in Princeton in honor of Einstein, 19 March 1949. We quote from Wigner [1967], p. 5 and p. 8.

[70] Wigner [1967], p. 8, note 4.

[71] Wigner [1967], p. 15. This fact was in reality known to several predecessors of Hamel, as we have shown in Chap. 1.

[72] Havas [1973] refers to Wigner [1954][1967] and gives several examples of this situation.

[73] This text is the lecture that Wigner delivered on 12 December 1963 upon receiving the Nobel Prize in physics that was conferred in recognition in particular of his "[. . .] discovery and application of fundamental symmetry principles."

Herglotz [1911], Engel [1916] and Bessel-Hagen [1921],[74] but without distinguishing the greater importance of her contribution.

In a long article written in collaboration with R. M. F. Houtappel and Hendrik van Dam, "The conceptual basis and use of the geometric invariance principles" (Houtappel [1965]), Wigner in note 20 on p. 606 informs the reader that the historical references were obtained thanks to Eugene Guth,[75] and that they are Jacobi [1866], Schütz [1897], Hamel [1904a] and, finally, Herglotz [1911]. Wigner and his co-authors continued, "F. Klein called attention to Herglotz work and encouraged F. Engel, E. Noether and E. Bessel-Hagen to make these ideas more explicit," and supplied references to relevant papers by each of these mathematicians: Engel's articles of 1916 and 1917, Noether's [1918c] and Bessel-Hagen's [1921]. Once again, Noether is not distinguished from her predecessors and her follower. The authors concluded their note with a short remark, "A more modern treatment of the subject was given by E. L. Hill [1951]," that may have discouraged generations of physicists from going back, beyond Hill,[76] to those original sources and, in particular, to Noether's article.

In still another text with the same title as the essay in *Symmetries and Reflections* described above, "Events, Laws of Nature, and Invariance Principles" [1995], written around 1980 but unpublished until 1995, Wigner again recognized a contribution by Noether and, before her, by Hamel, in the application of the principles of invariance in the formulation of conservation laws. And he added that, because the relation between invariance and conservation laws is much less evident in classical mechanics than in quantum mechanics, "the work of Hamel and Noether merits our respect" (p. 340) but apparently equally and no more than that.

To prepare his 1972 biographical article on Noether, Kimberling inquired from Wigner by mail regarding Noether, and he gave the following account of Wigner's reply:

> We physicists pay lip service to the great accomplishments of Emmy Noether, but we do not really use her work. Her contribution to physics that is most often quoted arose from a suggestion of Felix Klein. It concerns the conservation laws of physics, which she derived in a way which was at that time novel and should have excited physicists more than it did. However, most physicists know little else about her, even though many of us who have a marginal interest in mathematics have read much else by and about her.[77]

These remarks in fact suggest that, even though the subject of Noether's article had been central to Wigner's preoccupations since the 1920s, he had never read the

---

[74] Wigner [1967], p. 47, note 17.

[75] Guth (1905–1990) held a Ph. D. in theoretical physics from the University of Vienna, became a professor there until 1937, then taught at the University of Notre Dame in the United States and later worked at the Office of Naval Research. He is the author of an article [1970] in which he sought to show that the theory of Hilbert as a codiscoverer with Einstein of the equations of general relativity was a myth. This article was severely criticized by Mehra ([1974], p. 72, note 145, and p. 81, note 270) who held the opposite view. The most recent research contradicts Mehra's position, see, *supra*, Chap. 1, p. 41, note 48.

[76] See, *infra*, Chap. 4, p. 101.

[77] Kimberling [1972], p. 142.

original paper, although he did read Hill's rather inadequate treatment [1951] some time before 1965.

It is to Wigner's "Invariance in Physical Theory" [1949], and not to any scientific work of the first decades of the century, that C. N. Yang referred on 11 December 1957 in his lecture upon receiving, together with Tsung Dao Lee, the Nobel Prize in physics. In this influential discussion of "the general aspects of the role of the symmetry laws in physics," he wrote that "it is common knowledge today that in general a symmetry principle (or equivalently an invariance principle) generates a conservation law. For example, the invariance of the physical laws under space displacement has as a consequence the conservation of momentum, the invariance under space rotation has as a consequence the conservation of angular momentum. While the importance of these conservation laws was fully understood, their close relationship with the symmetry laws seemed not to have been clearly recognized until the beginning of the twentieth century." [78] Here Yang cited Wigner [1949] for references, with the result that he, too, entirely suppressed Noether's role.

Lanczos — Cornelius Lanczos (1893–1974)[79] was a *Privatdozent* in theoretical physics at the University of Frankfurt when he was invited by Einstein in 1928 to work with him in Berlin for a year. In 1931 he left Germany for Purdue University in Indiana, where he remained until 1946. He met Noether briefly around 1934 and, in an article concerning her contributions to the calculus of variations [1973], he outlined a vivid portrait of her remarkable personality, then furnished some biographical information, insisted on the importance to physics of "Noether's principle," i.e., the mathematical formulation of "Noether's theorem," added that the well-known book by Courant and Hilbert [1924] on this subject "is hard reading and does not come to grips with the real essence of the problem." Then he undertook to show that one can understand Noether's (first) theorem without recourse to group theory, which is at best paradoxical![80]

McShane — In the book edited by James W. Brewer and Martha K. Smith [1981] there is a six-page study of Noether's work on the calculus of variations by Edward James McShane[81] who "used to meet her in the early 1930s." McShane emphasizes the fact that "Emmy Noether's contribution [dealt with] both simple- and multiple-integral problems [...] invariant under a group of mappings of functions into functions." He mentions the possible influence of her "physicist brother" in leading her to this problem without specifying about which of Noether's three brothers he was

---

[78] Yang [1957], p. 95; 1964, p. 393; *Selected Papers*, p. 236.

[79] On Lanczos, see Sauer [2006] and Stachel [1994].

[80] On Courant and Hilbert [1924], see, *infra*, Chap. 4, p. 95. The successive editions of Lanczos's book on the calculus of variations [1949] will be discussed, *infra*, Chap. 5, p. 108.

[81] Pp. 125–130. The publications of McShane (1904–1989) were in the areas of the calculus of variations in several variables, integration theory, control theory and stochastic calculus. Accounts of his work have appeared in the *SIAM Journal of Control and Optimization*, 27 (1989), pp. 909–915, and in the *Notices of the American Mathematical Society*, 36 (1989), pp. 828–830. McShane was in Göttingen as assistant to Courant during the year 1932–1933, preparing the English translation of Courant's text on calculus (Mac Lane [1981], p. 67).

thinking.[82] He then describes "the case of simple integrands involving only the first derivative of the *n*-vector-valued function *y*," saying that this "case is still general enough to include all the conservation laws of physics" and, he adds, "in fact, I know of no book on the calculus of variations that indicates that 'Noether's theorem' goes beyond this." After a short summary of the proof of that "simple" case, McShane briefly gives an idea of the proof in the case of Lagrangians of higher order and then in that of multiple integrals. However, even he does not mention the case of infinitesimal symmetries that depend on derivatives that Noether had introduced. Finally, he states Noether's second theorem without analyzing it, and concludes by emphasizing the importance of these theorems. He then explains that they are no longer relevant to current research:

> Yet, they [Noether's two theorems] are ignored in many books on the calculus of variations and touched lightly in the others. They have fallen victim to a change in fashions. They constitute a major contribution to a highly formalistic aspect of the calculus of variations that received much attention in the nineteenth century, but was already becoming less interesting to analysts when her theorems appeared in Noether [1918c]. Problems involving multiple integrals and higher derivatives are largely ignored today, so it is quite natural that of her theorems, only the simple special case that we have presented still survives in the literature.

One must accept the validity of McShane's general history of the reception of Noether's theorems, but his judgment of the pertinence to the physics and mathematics communities of these theorems in their full generality was already inexact when he wrote because the formal calculus of variations had returned to the forefront of research with the articles of Israel M. Gel'fand and Leonid Dikiĭ (Dickey) [1975] and [1976], the second of which bears the title, in translation, "A Lie algebra structure in the formal calculus of variations," as well as the article by Gel'fand, Yuri I. Manin and Mikhail A. Shubin [1976] which also deals with the variational derivatives in the formal variational calculus, and the long article by Manin [1978] whose title is, again in translation, "Algebraic aspects of nonlinear differential equations." And McShane was writing nine years after the first article by Robert L. Anderson, Sukeyuki Kumei and Carl E. Wulfman [1972] who rediscovered Noether's concept of the generalized invariance of differential equations and its application to physics, and four years after Olver's article [1977] on evolution equations that possess infinitely many infinitesimal symmetries. Both articles make essential use of the infinitesimal method of Lie extended to the case of generalized infinitesimal transformations. It should thus have been clear to any well-informed observer that the "formal calculus of variations" coupled to Lie theory had indeed returned to style!

---

[82] See Kimberling's article in Brewer and Smith [1981], p. 5, where he mentions Alfred, a chemist, Fritz, who studied mathematics and physics in Erlangen and Munich, and Gustav Robert who is not known as a scientist. Fritz became a professor of applied mathematics, first at Karlsruhe and then at Breslau, before fleeing Nazi Germany to the Soviet Union. He was appointed a professor at Tomsk, in Siberia. He was arrested by the police 27 November 1937, and was executed 10 September 1941. See the article by his son, Herman D. Noether [1993]; also see Jacobson's note in Emmy Noether's *Gesammelte Abhandlungen / Collected Papers* [1983], p. 1, and Schlote [1991].

Wightman — In a note added in proof in footnote 166 to his article [1987] on Noether, Klein and Lie, Hans A. Kastrup (see, *infra*, p. 89) conveys the substance of a letter addressed to him by Arthur S. Wightman in January 1985 in which Wightman claimed that, "although it is true that theoretical physicists did not quote E. Noether'[s] paper in the fourtieth [*sic* for fourties], a number of them were quite aware of it."

The following testimony confirms the lack of a precise and direct knowledge of Noether's work among the best physicists as late as 1960.

Heisenberg — In the early 1960s, the philosopher of science Thomas S. Kuhn conducted a large number of interviews, including several with Werner Heisenberg (1901–1976), for the Archive for the History of Quantum Physics. Heisenberg had studied with Max Born in Göttingen in 1922–1923, where he then became a *Privatdozent*, and went to Copenhagen from 1924 to 1925 where he could discuss his research with Niels Bohr, Hendrik Kramers and John C. Slater. Thus arose the famous "BKS theory."[83] In the interview of 19 February 1963, Heisenberg recounts that the joint paper of these three leading physicists [1924] concluded "that energy was conserved only statistically," and that Pauli said "Well, that's too dangerous. There you try something which one shouldn't try." He then continues, "Much later, of course, the physicists recognized that the conservation laws and the group theoretical properties were the same. And therefore, if you touch the energy conservation, then it means that you touch the translation in time. And that, of course, nobody would have dared to touch. But at that time, this connection was not so clear. Well, it was apparently clear to Noether, but not for the average physicist. Also in Göttingen it was not clear. The Noether paper has been written in Göttingen, I understand. But it was not popular among the physicists, so I certainly wouldn't learn that from Born in Göttingen. By the way, do you recall when the Noether paper had been written? I think it must have been also around '23 or so." On Kuhn's candid answer "I've heard of that paper, but never looked at it," Heisenberg adds, "One really ought to look up the paper. I'm sure that the paper itself did not play a large role for the development of quantum theory. It did play a role for the development of general relativity. It was actually formulated in connection with general relativity, which was an interest with (Hilbert's) group and therefore also Noether. But it did not penetrate into

---

[83] Bohr (1885–1962), Kramers (1894–1952) and Slater (1900–1976) proposed their theory in order to reconcile the discontinuous, corpuscular aspect of light with its continuous, wave aspect. This theory did not yield conservation of energy for individual atomic processes, only statistical conservation for large numbers of processes. Dirac [1936] wrote, "Soon after the new theory was put forward, its predictions [...] were put to experimental test. The results [...] were unfavourable to the new theory, and supported conservation of energy. Shortly after that, the new quantum mechanics was discovered by Heisenberg and by Schrödinger, and was developed to provide an escape from the difficulties of the conflict between waves and particles without departing from conservation of energy. Thus the B.K.S. theory was found to be in disagreement with experiment and was no longer required by theoretical considerations, and it was therefore abandoned."

the circles of quantum theory, so I didn't realize the importance of that paper." The conversation then goes on discuss the importance of group theory for physics.[84]

From the silence of the physicists we deduce that, although Noether's results may have been known to some of them, in a general way, those results were not cited because none of them had actually read the article. Of course, some mathematicians and physicists were aware of Noether's work and cited it. We have one example. After the Second World War, at the University of Bonn, the mathematician Wolfgang Krull, who had studied with her in Göttingen in 1920–1921, "cited Noether both in his lectures on the calculus of variations and in his lessons on algebra."[85]

3.4 The Introduction to Noether's *Gesammelte Abhandlungen / Collected Papers*

The *Invariante Variationsprobleme* was of course reprinted in Noether's *Gesammelte Abhandlungen / Collected Papers* of 1983. For the introduction to this volume, written by Nathan Jacobson, the physicist Feza Gürsey (1921–1992) was invited to write the commentary on this article. As one would expect, he emphasized the importance of Noether's results, but in fact he only cited that of the first theorem and its application to quantum field theory, dealing with, on the one hand, the case of symmetry groups of the same type as that of translations and that of rotations, such as the Galilean group and the Poincaré group, and, on the other hand, the case of the group of diffeomorphisms, without distinguishing between them. Gürsey explained that to the infinitesimal symmetries of the action, indexed by α, there correspond by Noether's theorem the "charges," $Q^{(\alpha)}$, which are the integrals over space of the time-components $j_0^{(\alpha)}$ of the currents, i.e., of the conservation laws, $Q^{(\alpha)} = \int j_0^{(\alpha)} d^3x$, where $d^3x = dx\,dy\,dz$, and x, y and z are the space coordinates. The charges are thus conserved in the course of the time evolution described by the equations derived from the Lagrangian under consideration. "In quantum field theory, these charges are operators whose commutators satisfy the original Lie algebra [i.e., correspond to the Lie brackets of the Lie algebra of infinitesimal transformations], even if the action is not invariant under these transformations and [if consequently] the charges are not conserved. This is the basis of M[urray] Gell-Mann's current algebra,[86] of paramount importance in Particle physics." Gürsey then gave an expression for the Noether current associated with an internal symmetry which was "rediscovered half a century later by Gell-Mann." For independent variables x^μ and dependent variables u^a ($a = 1, \ldots, N$) and a first-order Lagrangian—this is

---

[84] Interview of Werner Heisenberg by T. S. Kuhn, Max Planck Institute, Munich, Germany, on 19 February 1963, Niels Bohr Library & Archives, American Institute of Physics, College Park, MD, USA, http://www.aip.org/history/ohilist/4661_6.html. I thank Michel Janssen for calling my attention to this testimony.

[85] Letter of Dr. Henri Besson, 15 August 2005.

[86] For an introduction to the algebra of currents, see, e.g., Treiman, Jackiw and Gross [1972].

what Gürsey assumed without calling attention to the fact that he had thus considered only a special case of Noether's result—, the current associated to a symmetry
$X = \sum_{a=1}^{N} X^a \frac{\partial}{\partial u^a}$, which he assumed to be both classical and vertical, may be written

$$J^\mu = \sum_{a=1}^{N} \frac{\partial L}{\partial (\partial_\mu u^a)} X^a.$$

In conclusion, Gürsey explained that "Emmy Noether was led to her great discovery
after being motivated by a physics problem, the invariance properties of the action
in General Relativity. She had then become interested in Hilbert's discovery of the
Lagrangian formulation of Einstein's theory." He further observes that Noether wrote two articles in 1918, the first [1918b] dealing with "the invariance properties and
the associated conserved quantities of systems of differential equations," and the
second [1918c] dealing with the formulation of dynamical systems and field theories from an action principle, and utilizing variational methods, and he concludes
by remarking, "The second paper also contains the expression for the current divergences and the construction of conserved charges (integral invariants) in the exact
symmetry case."

Here then is still another text which, while admiring Noether's important results
in mathematical physics and admitting that her "simple and profound mathematical
formulation did much to demystify physics," still remains an extremely incomplete
account of her results which were readily available while Gürsey was preparing his
commentary. One can expect more in a commentary, written as late as 1983, on
Noether's only text with direct bearing on physics.

3.5 Translations of the *Invariante Variationsprobleme*

There exists a Russian translation of the *Invariante Variationsprobleme* as well as
of Noether [1918b] by D. V. Jarkov in a book by Lev Solomonovich Polak [1959][87]
which is a large collection of Russian translations, with commentaries, of texts concerning the variational principles of mechanics, from Pierre de Fermat to Dirac. This
collection also contains a translation of Hilbert's article [1915].

An English translation of the *Invariante Variationsprobleme* was published in
1971 by Morton A. Tavel in the journal *Transport Theory and Statistical Physics*,[88]
preceded by a short commentary on the utility of the first theorem in the domains of
both classical and quantum mechanics, and followed by an application of Noether's
method to the case of a Lagrangian depending on the field variables by means of

---

[87] The translation of Noether [1918b] appears on pp. 604–610 and that of Noether [1918c] on
pp. 611–630.
[88] Noether [1971]. Tavel defended a doctoral thesis in physics at Yeshiva University in 1964.

integro-differential operators.[89] This translation, to the best of our knowledge the only one available in English before the one that introduces this study, suffers from slight misunderstandings of Noether's text, and seems to have had relatively modest diffusion[90] until Olver cited it in *Applications of Lie Groups to Differential Equations* [1986a], but it has been cited fairly frequently since then. In particular, it was mentioned by James D. Stasheff [1997], and by Pierre Deligne and Daniel Freed in their contribution, "Classical Field Theory" [1999], to the monumental treatise, *Quantum Fields and Strings: A Course for Mathematicians*, which was the result of a year's program in Princeton on quantum field theory.

The first French translation appeared in the French edition of the present book in 2004, and that translation was revised for the second edition in 2006.

There is a still unpublished Italian translation that was prepared in the early 1960s for Professor Enzo Tonti who, between 1960 and 1990, published research on variational principles and differential equations.[91]

3.6 Historical Analyses

Aside from the publications by Dick [1970] and then by Kimberling [1972] [1981] that are mainly biographical, there are several articles that deal chiefly or in part with the history of the Noether theorems. The earliest ones we have found are in Russian, because the historians of mathematics in the Soviet Union had become aware of Noether's work on variational problems well before their colleagues elsewhere took notice of it.

Polak 1959 — In addition to publishing a translation of Noether's article, Polak [1959] offers a very brief analysis of her contribution to the theory of variational principles in classical mechanics.[92]

Vizgin 1972, Polak and Vizgin 1979, Vizgin 1985 — Vladimir Pavlovich Vizgin's monograph [1972] is a study of the development of mechanics from antiquity to Lagrange, Hamilton and Lie, of the methods utilized at the end of the nineteenth century such as that of cyclic (also called ignorable or absent) coordinates due to Routh and Helmholtz, and of the development of the special theory of relativity, then of the work of Einstein, Hilbert, Lorentz, Weyl and Noether between 1913 and 1918 on the general theory of relativity. The chronology of these developments is summarized in a table, p. 153. The Noether theorems are analyzed in chapter 5; then, in the sixth and last chapter, Vizgin studies the diffusion of Noether's results in the literature of mathematical physics from 1951 to 1970, offering a bibliography

---

[89] Tavel [1971a, b].

[90] Tavel's translation was mentioned by Logan [1977]. It is now accessible at the site http://www.physics.ucla.edu/~cwp/pubs/noether.trans/english/mort186.html.

[91] We thank Franco Magri for calling our attention to this translation. Magri used it in an important article in Italian [1978] on the first Noether theorem. See, *infra*, Chap. 7, p. 141.

[92] Pp. 863–864 and pp. 911–913, note 213.

comprising 44 books and articles. This bibliography remains useful because certain items mentioned in it will not figure in this study. After publishing an article with Polak [1979] on the history of the first Noether theorem in physics, Vizgin published a book [1985] on the history of field theories in the first third of the twentieth century which would be translated into English in 1994. In its second chapter he analyzes Hilbert's theory and observes that his 1915 result concerning the conservation laws of general relativity is "a special case of Noether's second theorem, proved two and a half years later by Emmy Noether, who had come to Göttingen."[93] He then cites in a note the formulation of the second theorem with a reference to p. 239 in Noether [1918c], and refers to his earlier monograph [1972] on the subject of conservation laws. He asserts that this result has become classical, and that Pauli's encyclopedia article [1921][94] is proof of this fact, and he refers to Andrzej Trautman[95] for an interpretation of Hilbert's result using the Bianchi identities.

Mehra 1974 — The subject of Jagdish Mehra's study [1974] is the origin of the general theory of relativity. In his section 3.3, p. 20, "Space-Time and Invariants," and in the notes that accompany it he provides a detailed analysis of the work of Noether's precursors, Jacobi, Helmholtz and Hamel, and of her contemporaries, notably Engel, as well as of the content and importance of her two theorems.

Kastrup 1987 — Four years after the publication of Noether's *Gesammelte Abhandlungen / Collected Papers*, there appeared a long and well documented historical article on "The contributions of Emmy Noether, Felix Klein and Sophus Lie to the modern concept of symmetries in physical systems" by Hans A. Kastrup [1987], followed by a transcription of a discussion in which Henri Bacry, Louis Michel and Eugene Wigner, among others, participated. Kastrup[96] analyzes, on the one hand, the research that preceded Noether's in the area of differential invariants, some of which, according to him, made notable progress in understanding the relation between symmetries and conservation laws, and, on the other, the diffusion of Noether's results within the scientific community. We will not repeat here all the elements of his very informative history and interesting analyses, from which we have greatly profited, but it still must be remarked that even in an article so obviously based on a careful study of the texts, Kastrup could, in its section entitled "Emmy Noether's two theorems," reduce the statement of the first theorem to the case of first-order Lagrangians and classical symmetries, i.e., independent of derivatives, and that, in what follows, Kastrup does not mention more than a single article which referred to the second theorem, that of the relativists Peter G. Bergmann and Robb Thomson [1953]. Below, we shall show the importance of Noether's second theorem in the literature on general relativity from 1950 on.[97]

---

[93] Vizgin [1985], pp. 58–59 and again several times.

[94] See, *infra*, Chap. 4, p. 93.

[95] See, *infra*, Chap. 5, p. 110, and Chap. 6, p.126.

[96] Kastrup is also the author of a very complete monograph [1983] on the Hamilton–Jacobi theory and the De Donder and Weyl formalisms in the Lagrangian theory of dynamical systems.

[97] See, *infra*, Chap. 6, p. 123.

Olver 1986 — Peter Olver's book [1986a] on the application of Lie group theory to the study of differential equations, re-issued with additions in 1993, reached a wide audience. It constitutes a most valuable exposition of the Noether theory, including the second theorem, and its subsequent developments to which Olver had contributed. Furthermore, it contains, for each of the themes it treats, much historical and bibliographical information brought together in appendices to each chapter.

Rowe 1999 — The historical article by David Rowe [1999] contains a detailed study of the intellectual exchanges between Göttingen and Berlin, and of the interactions among Klein, Hilbert, Weyl and Einstein in the years 1915 to 1918, the period that corresponds to the creation and early diffusion of the general theory of relativity, as well as of Noether's role in the resolution of problems posed by that theory. The present commentary overlaps his in part, and some of the considerations which were dealt with in depth in his scholarly article are only sketched here.

Teicher 1999 — On 24 December 1991, the Mathematics Institute of Bar-Ilan University in Ramat-Gan (Israel) was officially named "The Emmy Noether Research Institute of Mathematics" and a conference commemorating her mathematical heritage was held in 1996. The *Proceedings* were edited by Mina Teicher and published three years later [1999]. This volume contains, in addition to a study by Nina Byers of the *Invariante variationsproblem* [*sic*] (cited, *supra*, p. 65), an article by Yuval Ne'eman [1999] on the applications of Noether's theorems to particle physics and to gauge theories.

It is clear from the above that while a small number of mathematicians, physicists and historians of science have been drawn to the history of this aspect of Noether's work since Polak and Vizgin, few publications have dealt with its diffusion, in part no doubt because it was so limited.

Chapter 4
The Transmission of Noether's Ideas, from Bessel-Hagen to Hill, 1921–1951

For some thirty years after the publication of Noether's *Invariante Variationsproble-me*, this work disappeared from the consciousness of all but a handful of writers of mathematics, mechanics and theoretical physics, while it would slowly re-emerge in later years. Three notable exceptions on which we comment below are the article by Erich Bessel-Hagen [1921], the treatise on the theory of invariants by Weitzenböck [1923], and the handbook of Courant and Hilbert [1924], after which we found only two articles that cite Noether, both in the literature of the 1930s on quantum mechanics. Then, Hill's article of 1951 initiated the diffusion of Noether's results, but provided only a summary of a very restricted case of her first theorem.

4.1 Bessel-Hagen and Symmetries up to Divergence

In 1921, Erich Bessel-Hagen (1898–1946) published an article in the *Mathematische Annalen* on conservation laws in electrodynamics, in which he determined in particular those which were the result of the conformal invariance of Maxwell's equations.[1] At the very beginning of the article, he explained that, on the occasion of a colloquium in the winter of 1920, Klein had posed the problem of "the application to Maxwell's equations of the theorems stated by Miss Emmy Noether about two years ago regarding the invariant variational problems."[2] In his introduction,

---

[1] The discovery of the conformal invariance of Maxwell's equations is generally attributed to Harry Bateman [1910] and to Ebenezer Cunningham [1910]. A half-century later, in a classic paper on conformal invariance, Thomas Fulton, Fritz Rohrlich and Louis Witten [1962] wrote, "According to Noether's theorem [...]" (p. 454) but gave no reference.

[2] "[...] das Herr Geheimrat F. Klein [...] abhielt, äußerte er den Wunsch, es möchten doch die vor etwa zwei Jahren Fräulein Emmy Noether aufgestellten Sätze über invariante Variationsproble-me auf die Maxwellschen Gleichungen angewandt werden" (Bessel-Hagen [1921], p. 258). After Klein had read Bateman's article [1910], he saw in the question of the role of the conformal group an exciting illustration of his Erlangen program, and he had the idea, as early as 1917, of applying the methods of group theory to the question of the invariance of Maxwell's equations under inversion, but Einstein, in a letter of 21 April 1917, discouraged him from pursuing that line of

Y. Kosmann-Schwarzbach, *The Noether Theorems*, Sources and Studies in the History of Mathematics and Physical Sciences, DOI 10.1007/978-0-387-87868-3_5, © Springer Science+Business Media, LLC 2011

Bessel-Hagen gives a summary of those two theorems, with their reference, Noether [1918c], and he restates their formulation in his first section, "The E. Noether theorems" ("Die E. Noetherschen Sätze"). He writes that he will formulate these two theorems slightly more generally than they were formulated in the article he cites, but that he "owe[s] these to an oral communication by Miss Emmy Noether herself."[3] This sentence in Bessel-Hagen's article has rarely been remarked in the secondary literature and, as a result, the few authors who sought a precise attribution of the invention of "symmetries up to divergence," a concept that simplified the formulation of Noether's results while permitting the inclusion of a still more general type of invariance than the one she studied in 1918, attributed it to Bessel-Hagen alone.[4] In fact, the infinitesimal transformations considered by Bessel-Hagen are required to satisfy, instead of the condition $\delta(f dx) = 0$, the weaker condition $\delta(f dx) = \mathrm{Div}\, C$, where C is a vectorial expression.[5]

Noether's fundamental equation (12) of [1918c][6] remains valid under this weaker assumption, provided that $B = A - f. \, \Delta x$ is replaced by $B = A + C - f. \, \Delta x$.

Bessel-Hagen ends his article by expressing his thanks to Noether and to Prof. Paul Hertz[7] "for the benevolent interest with which they supported [him] during the

---

investigation (*Collected Papers* 8A, no. 328, pp. 435–437; 8 (English), p. 318). See Rowe [1999], p. 211. Pursuing this idea over Einstein's objections, he persuaded Bessel-Hagen to work on this problem and its consequences for physics. The action of the conformal group on the equations of physics would not return to the fore as a subject for research until much later, after 1935, when it was studied by Oswald Veblen, by Dirac, and by Schouten and Johannes Haantjes. Nowadays, conformal field theory—a relativley young theory, dating to the 1980s—is extensivesly studied.

[3] "Zuerst gebe ich die beiden E. Noetherschen Sätze an, und zwar in einer etwas allgemeineren Fassung als sie in der zitierten Note stehen. Ich verdanke diese einer mündlichen Mitteilung von Fräulein Emmy Noether selbst," Bessel-Hagen [1921], p. 260.

[4] This attribution can be found in Olver [1986a], pp. 288 and 367, and in Bluman and Kumei [1989], p. 275. While Kastrup [1987] did not speak about symmetries up to divergence, Mehra, for his part, writes no more than this misleading sentence, "Her work was carried on, or explained better, in terms of physics, by Bessel-Hagen" (Mehra [1974], p. 22). But Stefan Drobot and Adam Rybarski, in an article [1958] where they applied the Noether theorems to problems in hydrodynamics, did observe (p. 404), before citing Bessel-Hagen's article, that "the idea of div-invariance was introduced by E. Noether," while Peter Havas [1973], note 7, also recognized that the generalization of Noether's first theorem formulated in Bessel-Hagen's article was in fact due to Noether herself. Rowe [1999] does not mention specifically the advance constituted by the passage from infinitesimal symmetries to infinitesimal symmetries up to divergence, but he informs us (note 17) that "Bessel-Hagen consulted with Emmy Noether while writing [this] paper," which seems a bit weaker than what Bessel-Hagen's admitted "debt" to Noether implies.

[5] The condition in Noether's article, $\delta(f dx) = 0$, means the invariance of the Lagrangian f, while the condition $\delta(f dx) = \mathrm{Div}\, C$ means the invariance of the action integral $\int f dx$.

[6] See, *supra*, p. 8.

[7] Paul Hertz (1881–1940), a theoretical physicist and mathematician. According to Mac Lane [1981], p. 75, Hertz was an *Ausserordentliche Professor* in mathematics in Göttingen in 1933. He succeeded in leaving Germany before the war, settled in the United States, and died in Philadelphia. In 1910 Hertz had pointed out to Einstein that a statement in his paper on the second law of thermodynamics (1903) needed to be clarified and proved (see Pais [1982], p. 67) and, in 1914–1915, he helped Hilbert understand an argument of Einstein (see Rowe [2001], p. 415). In a letter to Fischer, written in November 1915, a few months after her arrival in Göttingen, Noether mentions the

execution [of his work]."[8] However, Weyl does not mention Noether in his summary of Bessel-Hagen's article for the *Jahrbuch über die Fortschritte der Mathematik*,[9] and Pauli, in the famous encyclopedia article on the theory of relativity [1921], about which more below, thanks Bessel-Hagen but omits mentioning Noether altogether. Both omissions are, to say the least, surprising.

4.2 Pauli 1921 and 1941

Shortly after the publication of Noether's article, Wolfgang Pauli (1900–1958), then a young but already brilliant physicist, wrote the article *Relativitätstheorie* for the *Encyklopädie der mathematischen Wissenschaften* [1921] in which he thanks Klein and also—as said above—Bessel-Hagen who had read some pages of the printer's proofs, but Noether is cited only once, in a footnote on p. 48 where he refers to [1918b], and that reference does not concern the question of the conservation of energy in general relativity, only the search for invariants in Riemannian geometry. (Another note refers to F[ritz] Noether, while the index erroneously refers to E. Noether twice.) The lack of any reference to the *Invariante Variationsprobleme* in Pauli's article is all the more surprising because a letter of 8 March 1921 from Klein to Pauli[10] informs us that he had lent Pauli his personal notes for the preparation of Pauli's long report [1921] and that he had called his attention to "the theorems of Noether," without supplying a precise reference as though they were so well known that identifying them was not necessary, and to Bessel-Hagen's as yet unpublished results that he, as editor of the journal, planned to publish in the *Mathematische Annalen* as soon as the physicists, Planck and Pauli himself, would give him a favorable opinion regarding the physical significance of the conservation laws that Bessel-Hagen had obtained.

Twenty years later, after the Solvay Congress of 1939 was canceled because of the war, Pauli published the gist of his intended lecture in a much read article in the *Reviews of Modern Physics* [1941]. In the first part, he considered the field equations derived from a Lagrangian and stated the "continuity equations" for both translation-invariant and Lorentz-invariant Lagrangians, but he did not refer to Noether's general method, while in the second part he studied the field theories for spin 0, 1/2 and 1, and their applications. He defined the "canonical" energy-momentum tensor, then, following work of Frederik J. Belinfante [1939] [1940] and Léon Rosenfeld

---

presence of "the physicist Hertz" who has studied invariant theory in Gordan [1885] ("[...] sogar der Physiker Hertz studiert Gordan–Kerschensteiner"), letter quoted by Kastrup [1987], p. 122.

[8] "[...] für ihr wohlwollendes Interesse, mit dem sie mich bei der Durchführung unterstützten, zum Ausdruck zu bringen," Bessel-Hagen [1921], p. 276.

[9] Vol. 48 (1921–1922), p. 877.

[10] Pauli [1979], pp. 27–28. See a translation of this letter, *infra*, in Appendix III, pp. 159–160. This letter has been cited by Rowe [1999], p. 228. Klein's letter is particularly interesting for our discussion, and also because it gives his views on the question of the Poincaré–Einstein and Hilbert–Einstein priorities.

[1940], a symmetric tensor.[11] He referred to his own encyclopedia article [1921], and then cited Hilbert [1915] and Weyl [1929], but not Noether.

4.3 Weitzenböck 1923

The Austrian mathematician Roland Weitzenböck (1885–1955), who had obtained his doctorate in Vienna in 1910, taught in Graz, then in Prague and, in 1923, was appointed professor in Amsterdam where he remained. It was in the Netherlands that his book on the theory of invariants [1923] which we cited above[12] appeared. In this 408-page book, he cites Noether several times in the part that deals with algebraic invariants. (In fact, he refers to some of her articles, not by their date of publication, but by the year of their submission, identifying [1915] correctly enough but also citing [1911] as having been published in 1910 and [1916b] as dating from 1915.) In chapter 13, on the invariants of differential forms, in a section entitled "Emmy Noether's Reduction Theorem" ("Der Reduktionssatz von Emmy Noether"), he proves one of Noether's results of [1918b]. But the chapter that interests us the most is chapter 14, on integral invariants of differential forms.[13] Given a homogeneous differential form and a scalar function depending on the coefficients of that form and their derivatives up to a given order, Weitzenböck considers the integral of that function on a prescribed domain and calls it an integral invariant if its value is independent of the coordinate system that was chosen. He then deals with the variational problem posed by the determination of the extrema of an integral invariant with fixed boundary conditions when the coefficients of the differential form vary. He determines the related Euler–Lagrange equations, which he simply calls the "field equations" (*Feldgleichungen*), and he shows that the invariance of a Lagrangian with respect to the continuous group of coordinate transformations gives rise to conservation laws. To accomplish these calculations, he adopts the method of Lie and replaces invariance under the group action by invariance with respect to infinitesimal deformations. These methods are very close to those of Noether in [1918c], and Weitzenböck refers to that article in a note (p. 369). He concludes his study of conservation laws by citing the fourth edition of Weyl [1918b], and references therein, specifying Klein [1918b] and also Noether [1918c], but not Bessel-Hagen.[14] Finally, he shows how to derive Maxwell's equations as well as Einstein's gravitational equations from a variational principle, but he does not speak about the conservation laws associated with these theories. Later, in 1929, Weitzenböck corresponded with Einstein, pointing out to him that the components of the connection that Einstein considered in several notes in 1928 and 1929 had already been pub-

---

[11] For Belinfante and Rosenfeld, see, *infra*, p. 96.

[12] See Chap. 1, p. 30, note 3.

[13] Pp. 363 ff.

[14] "Zu obiger Ableitung vgl. H. Weyl, Raum, Zeit, Materie 4 Aufl., S. 211 ff. und die dort genannte Literatur, von der besonders angeführt sei: F. Klein, Göttinger Nachr. 19. Juli 1918 und E. Noether, Ebenda 26. Juli 1918," Weitzenböck [1923], p. 377.

lished six years earlier in his article in the *Encyclopädie* [1922] and again in his book [1923].[15]

Study,[16] in his treatise "Introduction to the Theory of Invariant Linear Transformations Based on the Vector Calculus. Part I" [1923] (no Part II seems to have been published) which was exactly contemporaneous with Weitzenböck's book, covered much the same material concerning invariant forms in three variables. He referred to Gordan, with whom he had studied invariant theory in Erlangen back in 1887.[17] Weitzenböck cites Study in various contexts and Study refers to Weitzenböck's earlier work but, since he does not deal with differential invariants, he never refers to Noether.

4.4 Courant and Hilbert 1924

Paragraph 11 of section 10 in chapter 4 of Courant and Hilbert [1924], "Complements and exercises on chapter 4" ("Ergänzungen und Aufgaben zum vierten Kapitel") is entitled "The theorems of E. Noether on invariant variational problems. Integrals in particle mechanics" ("Die Sätze von E. Noether über invariante Variationsprobleme. Integrale in der Punktmechanik").[18] We observe that "the theorems of E. Noether" of the German text became "E. Noether's theorem," in the singular, in the English translation of 1953 and later editions, a minuscule but telling change. All the editions include a reference to Noether [1918c] in a footnote. Courant[19] and Hilbert treat the case of variational problems with two independent variables and one or two dependent variables, and restrict their discussion to the case of classical symmetries. They further indicate that in fact the results can easily be extended to the case of an arbitrary number of independent and dependent variables, and finally they apply the result of the first theorem to the mechanics of a system of material points for which they derive the conservation of the linear momentum and of the angular momentum from the invariance of the Lagrangian under translations and rotations but, despite the title of the section, they refer to Bessel-Hagen [1921] rather than to Noether for the conservation of energy.[20] Then they sketch Noether's second theorem for the case of a symmetry that depends on an arbitrary function p and its

---

[15] See section 3.2 of Sauer [2006].

[16] See Chap. 3, p. 71, note 25.

[17] See Hawkins [2000], p. 236. Hawkins discusses Study's work at length and stresses the connections between Study's book and Weyl's work.

[18] This paragraph became paragraph 9 of section 11 in the 1931 edition, and paragraph 8 of section 12 in the subsequent editions. All the literature cited for this chapter is earlier than 1913, except for Tonelli [1921], in which Noether's work does not appear.

[19] Richard Courant (1888–1972) had been Hilbert's student. He was appointed to a professorship at Göttingen in 1922. Forced to leave Germany in 1933, he eventually settled in New York, where he founded the prestigious mathematics institute that bears his name at New York University.

[20] P. 264 of the English translation (1953).

derivatives up to order k, "i.e., the Euler equations are not mutually independent."[21]
As an example they treat the homogeneous case that had already been treated by
Noether where one finds the identity $\dot{x}\dfrac{\delta L}{\delta x} + \dot{y}\dfrac{\delta L}{\delta y} = 0$, and they conclude with a
footnote, "For a more detailed discussion and for generalizations and applications
to mechanics, electrodynamics and relativity theory, see the paper by E. Noether
referred to above and the references given there."[22]

4.5 In Quantum Mechanics

Rosenfeld 1930, 1940 — It appears that the earliest reference to Noether in an article by a physicist is the application of Noether's theorems to quantization problems
by Léon Rosenfeld (1904–1974) in an article in German published in the *Annalen
der Physik* [1930]. The paper by Rosenfeld, who was then in the physics department
of the Eidgenössische Technische Hochschule (ETH) in Zurich, may be the first attempt to construct a quantum theory of the gravitational field. He recalls the result
of the classical theory, and cites Noether [1918c] on this question,[23] then proposes a
method of quantization, passage from the c-functions to the q-functions,[24] introducing, for each variable, the half-sum of that variable and its adjoint matrix. Further
on, he extends Noether's result concerning the action of symmetries on the conservation laws to the case of noncommuting variables, and, on this question (p. 134), he
cites Section 5 of her article. A reference to Rosenfeld's article is made in the first
of the two papers by Komorowski [1968] dealing with Noether's first and second
theorems.

In a later article [1940], written in French this time, during his visit to the Institute of Advanced Study at Princeton, Rosenfeld discusses the determination of
the energy-momentum tensor in general relativity and in several other field theories with both tensorial and spinorial variables, obtaining results in agreement with
those derived independently by Belinfante [1939]. Both articles would soon be cited
by Pauli [1941]. Rosenfeld states that he will use "the fundamental identites, which
result from the invariance of the Lagrangian, $\int \mathscr{L}dw$, under arbitrary coordinate
transformations. Though these matters are well known, we shall quickly review the
derivation of these identities [...]."[25] The bibliography furnished by Rosenfeld contains the articles we have cited above by Lorentz [1915] [1916] and Hilbert [1915],

---

[21] "d. h. die Eulerschen Gleichungen sind nicht unabhängig voneinander," Courant and Hilbert
[1924], p. 218, 2nd ed., 1931, p. 226; English translation, 1953, p. 266.

[22] "Betreffs genauerer Angaben, Verallgemeinerungen und Anwendungen in der Mechanik, Elektrodynamik und Relativitätstheorie vergleiche man den oben genannten Aufsatz von E. Noether
und die dort angegebenen Arbeiten," *ibid.*

[23] The reference on his p. 119 is to p. 211, *sic* for 241, of Noether's article.

[24] "Übergang zu den q-zahlen" is the title of his section on quantization.

[25] "[...] en tirant parti des identités fondamentales qui résultent de l'invariance de la fonction
de Lagrange $\int \mathscr{L}dw$ pour une transformation quelconque des coordonnées. Quoiqu'il s'agisse

then a large number of papers by De Donder from 1916 to 1924 and his book [1935] (see, *infra*, p. 100). He also cites Weyl, Pauli and Fock. In addition, there are two references to Klein, the first is to his [1918a] and the second is to "*Gött. Nachr.* 1918, p. 235," a curious *lapsus calami* since p. 235 is the first page of Noether's *Invariante variationsprobleme*.

Belinfante 1939, 1940 — Frederik J. Belinfante (1913–1991) is the author of a series of papers in English on the theory of higher-order spinors which he called "undors." These papers were published in 1939 and 1940, shortly after he completed his thesis at the University of Leiden under the direction of Kramers. In the first of two articles in *Physica* [1939], he determines the analogue of angular momentum for mesons, particles of spin 0 represented by wave functions which are spinors with two indices, and in its sequel [1940], he discusses the energy density as well as the densities of linear and angular momenta for fields on general relativisitc space-times, including spinor fields, obeying field equations derived from a Lagrangian, and he shows how the modification of the Lagrangian by the addition of a divergence influences the quantities thus determined. He refers to Pauli [1921], to his previous publications, in particular to [1939], and to Rosenfeld [1940], then in press, but not to Noether.

Markow 1936 — The physicist Moisei A. Markow (1908–1994) was a student of V. A. Fock (1898–1974) and of Georg B. Rumer (1901–1985) who had been an assistant to Max Born in Göttingen from 1929 to 1932. Markow was a specialist in quantum mechanics who became the director of the Moscow theoretical optics laboratory. His article [1936] on Dirac's theory of the electron[26] was published in German in the *Physikalische Zeitschrift der Sowjetunion* when he was a member of the physics institute of the U.S.S.R. Academy of Sciences in Moscow. It begins, "For the group G_{10} one obtains, with the aid of Noether's theorems, 10 conservation laws ($\mathrm{div}\,A = T$)."[27] In his introduction Markow refers to "the well-known theorems of Noether," then states that it is with the aid of these theorems, based on an action principle and group theory, that he derives the conservation laws in his article.[28] After having recalled Noether's (first) theorem with the reference to her article on p. 235 of the *Gött. Nachr.*, and distinguishing the case of the groups with a finite number of parameters from the case of the "infinite" groups, he applies the method of Noether and Bessel-Hagen, whom he also cites (on pp. 777 and 779), to the determination of the conservation laws of the Dirac equation[29] which in fact derives from a Lagrangian invariant under the action of the 10-parameter (inhomogeneous) Lorentz

---

toujours de choses bien connues, nous rappellerons rapidement la déduction des identités [...]," Rosenfeld [1940], p. 5; English translation, p. 714.

[26] Dirac's article had appeared in the *Proceedings of the Royal Society of London* in 1928.

[27] "Für die Gruppe G_{10} wurden mit Hilfe der Lehrsätze von Noether 10 Erhaltungssätze abgeleitet ($\mathrm{div}\,A = T$)," Markow [1936], p. 773.

[28] "Die Erhaltungssätze werden in der vorliegenden Arbeit mit Hilfe der bekannten Lehrsätze von E. Noether abgeleitet (Wirkungsprinzip, Gruppentheorie)," p. 774.

[29] "If one applies the known theorems of Noether to the Lagrangian expression of Dirac's equations, then one can obtain all the invariants of the form $\mathrm{div}\,A = T$" ("Wendet man die bekannten Lehrsätze von Noether auf den Lagrangeschen Ausdrück der Diracschen Gleichungen an, so kann man alle Invarianten von der Form $\mathrm{div}\,A = T$ erhalten," pp. 775–776).

group, i.e., the Poincaré group. The references to Noether are numerous (pp. 773–776, 789, 798); then, in connection with covariance in general relativity, he cites the closely related article by Klein [1918b]. In this paper, Markow emphasizes the usefulness of Noether's results and cites them prominently. Finally, he thanks Prof. G. B. Rumer for his encouragement.[30] However, it is not obvious that Markow's knowledge of Noether's work was transmitted by Rumer. In fact, Rumer, in an earlier article [1931b], which was written in Göttingen, proved the Lorentz invariance of the Dirac operator but did not allude to any associated conservation laws, while in his articles on the general theory of relativity in the *Göttinger Nachrichten* [1929] and [1931a] he cited Weyl but never Noether. Also in contrast with Markow's citation of Noether, Fock's treatise on "Space, Time and Gravitation" [1955][31] contains no reference to her article although it contains a discussion of conservation laws in both the "mechanics of point systems" and general relativity. It seems that none of the many articles written by Fock between 1922 and the 1950s refers to her work either.

4.6 Negative Results

The surveys of Vizgin [1972] and Kastrup [1987] and our own research have yielded a surprisingly small number of references to Noether before 1950. In particular, we are still astonished by the absence of citations dealing with invariance and its related mathematics in the corpus of the then standard but now classical textbooks on the variational calculus. There is nothing on invariance problems in the treatises of Tonelli [1921], Levi-Civita and Amaldi [1923], Bliss [1925],[32] Forsyth [1927], Ames and Murnaghan [1929] who treat cyclic coordinates but not general invariance properties, nor later, in that of Elsgolc [1952], originally written in Russian and translated into English in 1961, cited by Gel'fand and Fomin [1961], nor in Pars [1962].

As for the literature on quantum mechanics, Gregor Wentzel, then a professor at the University of Zurich, states in his book published in Vienna [1943] only that "it is well known that, as in classical mechanics, the validity of conservation laws is related to certain properties of invariance of the Hamiltonian function,"[33] but enters into no further details. Similarly, the second part [1957] of Wentzel's lectures at the

---

[30] "Für die Anregung zu dieser Arbeit."

[31] In his discussion of the problem of energy in general relativity in this treatise, which first appeared in Russian, then in English translation in 1959, Fock distinguishes clearly the work of the Einstein school in the period 1938–1954 from his own and that of his collaborators.

[32] In an address at a meeting of the American Mathematical Society in December 1919, entitled "Some recent developments in the calculus of variations," Gilbert Ames Bliss (1876–1951) chose to speak of "a new method of treating the second variation," but made no reference to other topics (Bliss [1920]).

[33] "Bekanntlich ist die Gültigkeit von Erhaltungssätzen (wie in der klassischen Mechanik) an bestimmte Invarianzeigenschaften der Hamiltonfunktion gebunden," Wentzel [1943], p 11; English translation, 1949, p. 11.

Tata Institute in Bombay contains long developments on continuity equations, but the only reference (p. 18) is to Belinfante's tensor [1939].

We also observe that Edward M. Corson studies the conservation laws of variational problems in his treatise [1953], cites Weyl and De Donde, but never Noether.

Cartan 1922 — It should be noted that Élie Cartan (1869–1951), in chapter 18 of his "Lectures on Integral Invariants" [1922], studies "the integral invariants and the calculus of variations," and, in particular, the case of a Hamiltonian independent of time, and then generalizes his results to the case where the Hamiltonian does not depend on one of the phase-space variables. Of course he cites Lie, who had published two articles on integral invariants in 1897,[34] but he mentions no results from the work of Noether, nor does he cite any member of the German school, though he does cite various French, Belgian, Italian, English and Polish publications.[35] The lack of references to the Göttingen school may have been due to the great difficulties of communication during the war years.[36]

Carathéodory 1935 — Constantin Carathéodory (1873–1950) should have been well acquainted with Noether in the period when she published her article because he had been a professor of mathematics at Göttingen since 1913 when he succeeded Klein, and one finds his name once or twice a year, from 1915 to 1918, on the lists of speakers at meetings of the Göttingen Mathematical Society.[37] On 30 November 1915 he and Hilbert presented the first part of a paper on the theory of invariants, and completed their communication on 7 December. On 23 January 1917 he lectured on the "Variational problems with symmetrical transversality." Noether did not allude to this communication. Conversely, Carathéodory's article [1929] in the *Acta* of Szeged on the calculus of variations in several variables which was to inspire Weyl [1935b] did not treat the question of invariants, so he had no reason to cite Noether's research in it, and his "Calculus of Variations and Partial Differential Equations of the First Order" [1935] contains nothing on symmetries in general nor on Noether's work in particular.

---

[34] Lie [1897a,b]. While, at the beginning of the first of these papers, Lie cites Cartan's article [1896] on the integrals associated to systems of lines or planes in space which remain invariant under the group of metric-preserving linear transformations, Cartan explains in this article (p. 176) that the expression "integral invariants" ("invariants intégraux") is due to Poincaré, and that in fact all the integral invariants he had discussed could be found by Lie's methods ("D'après les méthodes de M. Lie, on a le moyen de trouver tous ces *invariants intégraux*").

[35] It is curious that those are the victorious countries in the first World War or their allies, but it may also have been a coincidence because a rapid review of Cartan's articles shows that whatever their date, he gives very few bibliographical references, and in 1893 he had no scruple about publishing an article in German, despite the Prussian victory over France in 1871.

[36] See Reid [1970], pp. 144–145, for a discussion of the relations between the French and German scholars during the war and an account of the hardships of life in Göttingen. Even now it is hard to find in Paris certain issues of German mathematics journals from the war years because many libraries had suspended their subscriptions or were unable to obtain copies of the journals of their nation's adversaries.

[37] See Appendix V, p. 167.

De Donder 1935 — Between 1901 and 1935 Théophile De Donder (1872-1957) published almost two hundred papers concerning the calculus of variations and mathematical physics and, in particular, a number of articles on "Einsteinian gravity" ("la gravifique einsteinienne"), and in fact he continued to publish until 1955. In 1912 he had published a note in the *Comptes rendus* of the Paris Academy of Sciences, "On the invariants of the calculus of variations" [1912], where he introduced a condition for obtaining, by a change of variables, a Lagrangian which would not depend explicitly on the independent variable, but it does not appear that Noether was acquainted with that paper. On the other hand, when De Donder published a general formula in the *Bulletins de l'Académie Royale de Belgique* [1929] expressing the variation of a Lagrangian in terms of its variational derivative, his article lacked any reference to Noether even though the subject was so close to hers. In his subsequent treatise, "Invariant Theory of the Calculus of Variations" (*Théorie invariantive* [*sic*] *du calcul des variations*) [1935], one finds a study of the most general variational problems, as well as the definition of self-adjoint linear differential operators and their characterization as the variational derivatives of quadratic Lagrangians (chapter 15). These developments anticipate the introduction of adjoint operators by Magri and by Vinogradov.[38] The titles of De Donder's book and of Noether's article are strikingly similar. However, unlike Noether, De Donder does not deal with conservation laws of variational problems. In the title of his book and in the text, he uses the word "invariantive" to express the tensorial nature of the quantities he is considering. He was clearly familiar with the work of Einstein and he refers to his publications, but he only cites Hilbert once and refers to no other work of the Göttingen school.

De Donder corresponded with Einstein from June to August 1916 despite the fact that Belgium and Germany were in opposing camps during the war. Much later, in his book [1935], he shows that the gravitational field equations can be derived from a variational principle and, in a note, he refers to one of his articles that had appeared in the *Verslag Akademie Amsterdam* [1916] and then adds,

> This note, written in September 1915, could only be sent to Mr. H. A. Lorentz tardily because of the German occupation. The generalized Hamilton principle has also been utilized by D. Hilbert, *Göttinger Nachrichten* (November 1915). Compare with A. Einstein, *Berliner Berichte* (November and December 1915). In these notes, Einstein bases his research on considerations relative to invariance and to covariance, without mentioning the variational principle [invoked in the Amsterdam note and in Hilbert's article].[39]

A reading of this note reveals the genesis of a quarrel about scientific priority. In fact, Einstein would write on a variational formulation of general relativity only the

---

[38] See, *infra*, Chap. 7, p. 140.

[39] "Cette note, écrite en septembre 1915, n'a pu être envoyée que tardivement à M. H. A. Lorentz, à cause de l'occupation allemande. Le principe de Hamilton géenéralisé a aussi été utilisé par D. Hilbert, *Göttinger Nachrichten* (novembre 1915). Comparer à A. Einstein, *Berliner Berichte* (novembre et décembre 1915). Dans ces notes, Einstein base ses recherches sur des considérations relatives à l'invariance et à la covariance, sans mentionner le principe variationnel," De Donder [1935], p. 173.

following year[40] and publish an article [1916b] on this subject at the end of October, while De Donder's role in the development of the general theory of relativity has rarely been cited.[41] The reason may be that the war and its aftermath prevented the diffusion of the results of his research in the German universities.

Schouten 1951 — Schouten's book, *Tensor Analysis for Physicists* [1951], presents an account of the tensor calculus on manifolds and its applications to classical dynamics, to special and general relativity, and to "Dirac's matrix calculus." It contains a paragraph on "Special cases of first integrals," treated by the method of ignorable coordinates of Routh and Helmholtz, but no treatment of the link between first integrals and symmetries in general. If Schouten was aware of Noether's work, he refrained from presenting it in this introductory treatise. More probably, he had no knowledge of it.

4.7 Hill's 1951 Article

Edward Lee Hill,[42] in an eight-page article [1951], presents a synthesis of Hamilton's principle and the mathematics of conservation laws for students of mathematical physics. He proposes to give a simplified account of the theory while referring to Klein [1918b], to Noether [1918c] of course, to Bessel-Hagen [1921] and to De Donder [1935], but he never mentions Noether's second theorem and provides only a simplified version of the first theorem, "adapted to the needs of the student of mathematical physics." His version of the first theorem, restricted to first-order Lagrangians and classical symmetries, would be repeated in the literature very often. While he never mentions generalized symmetries, he does remark that the mathematical theory can be generalized to the case where the Lagrangian contains derivatives of arbitrary order, and he gives the above references. His article uses notation close to Noether's. At equation (12), for example, he introduces the vertical representatives which he denotes by $\delta_* \phi^\alpha$, where Noether used the notation $\bar{\delta} u_i$. As examples he discusses the cases of classical mechanics, the conservation of linear momentum, energy, angular momentum and the theorem of the center of mass, and, for field theory, he treats the case of the scalar meson, i.e., the conservation laws associated with the invariance of the Klein–Gordon equation[43] under the action of the

---

[40] See Chap. 1, p. 40, note 47.

[41] Notable exceptions are the monograph by Kastrup [1983] and the article by Catherine Goldstein and Jim Ritter [2003]. See also Rosenfeld [1940] and Hill [1951].

[42] Hill (1904–1974) was a physicist who taught at the University of Minnesota in Minneapolis. He was the coauthor, with E. C. Kemble, of the second in a series of two long articles on the principles of quantum mechanics that appeared in 1929 and 1930 in the first volumes of the *Reviews of Modern Physics*. These articles are cited in the English translation (1932) of Weyl's *Gruppentheorie und Quantenmechanik* [1928], p. 400, where both are, erroneously, attributed to Hill and Kemble.

[43] Unlike the equations of classical mechanics, this equation was formulated posterior to Noether's article, by Erwin Schrödinger, shortly before the publication of the articles by Walter Gordon and Oskar Klein in 1926 and 1927 in which it also appeared.

Poincaré group. The short review of Hill's article by Leopold Infeld in *Mathematical Reviews* fails to mention Noether's name.

It is this article by Hill which would be cited by physicists as their reference to Noether but, since it truncated her results and gave them only in a special case, it had the consequence that the general results of Noether remained unknown for several years even after it appeared.

Chapter 5
The Reception of Noether's First Theorem after 1950

In the inventories that we shall present in this chapter and the next, we shall try to show how the two theorems of Noether that interest us here were discussed within the community of mathematicians and physicists in the books and articles that were published between 1950 and 1980. We tried to perform as wide a survey as possible in order to identify references to the *Invariante Variationsprobleme* in that literature that would support our impression that, in contrast to the period between 1921 and 1950, this article was indeed cited but not yet very frequently, and almost always in the truncated form in which Hill had presented it. We found that Noether's name first appears in connection with differential invariants, the calculus of variations or the general theory of relativity in the books of Paul Funk [1962] and of Israel M. Gel'fand and Sergei V. Fomin [1961], which do not seem to be indebted to Hill. This is hardly surprising since Noether's German should not have been an obstacle for Funk who had direct access to her text and did not need the help of a well-intentioned intermediary like Hill, while the Russian mathematicians were still relatively isolated from scientific developments in the United States and would not, in most cases, have been aware of Hill's article. But, for France and the English-speaking world, despite Hill's article, Noether's first theorem was still not discussed until around 1965.

Of course our survey is not exhaustive, could not be exhaustive, but we think that we have looked at the most important texts of this period, more of them than we would cite here because negative historical results, however important, are not always interesting to read. Other reviews of this literature, such as Joe Rosen's [1972] and [1981], Vizgin's [1972] and Kastrup's [1987], complement our survey, and Vizgin and Kastrup have observed the same paucity of references to Noether in this period.

Y. Kosmann-Schwarzbach, *The Noether Theorems*, Sources and Studies in the History of Mathematics and Physical Sciences, DOI 10.1007/978-0-387-87868-3_6,
© Springer Science+Business Media, LLC 2011

5.1 Symmetries and Conservation Laws in Classical Mechanics and Quantum Physics

Not only did Noether's first theorem become the basis for most presentations of the relation betweeen symmetries and conservation laws in both classical mechanics and classical field theory, though in fact it was still not always cited in that context, but it also became fundamental for the relation between invariances and conservation properties in quantum mechanics and quantum field theory. As Gürsey argued in his comments on the *Invariante Variationsprobleme* in the introduction[1] to the *Gesammelte Abhandlungen / Collected Papers* of Noether, the algebra of currents in the work of Murray Gell-Mann, Julian Schwinger, Roman Jackiw and other physicists can be founded on Noether's principle even though the link to Noether was not evident to them in their earliest formulations.

5.2 On Some Encyclopedia Articles

Let us first remark that among the general scientific works that mention Noether, even those that devote much space to her work as an algebraist do not consider the *Invariante Variationsprobleme* worth citing among her more important papers.

One finds nothing concerning Noether's pre-1919 work in the long article by Oystein Ore in the 1967 edition of the *Encyclopædia Britannica*, although Ore gives a thorough account of her later work in algebra, nor in the *Micropædia* of the 2002 edition of the *Britannica*. In the 1974 edition of the *Dictionary of Scientific Biography*, Edna E. Kramer draws nearly word for word on Weyl's eulogy [1935a]:

> In 1915, Hilbert invited Emmy Noether to Göttingen. There [...] she applied her profound invariant-theoretic knowledge to the resolution of problems which he and Felix Klein were considering. [...] She was able to provide an elegant pure mathematical formulation for several concepts of Einstein's general theory of relativity.

But the *Invariante Variationsprobleme* does not figure in the list of Noether's principal publications that concludes the article.

However, in the 1998 edition of the *Brockhaus Enzyklopädie*, the notice, "Noether," does mention her work on the theory of invariants and differential invariants. It is followed by another, "noethersches Theorem," which is inaccurate but well-intentioned. While Noether's first theorem is mistakenly identified as "a theorem formulated by A. E. Noether and D. Hilbert," it is stated correctly and its importance in classical and quantum physics is stressed. This encyclopedia contains no mention of the second theorem.

In the early editions[2] of the French *Encyclopaedia Universalis* one can find a very generous article by Dubreil that treats Noether as the "creator of abstract algebra,"

---

[1] See, *supra*, Chap. 3, p. 86.

[2] 1968, 1973, 1980, 1989 editions.

but never mentions the *Invariante Variationsprobleme*.[3] By the 1990s, someone in the editorial department of the *Encyclopaedia Universalis* seems to have realized that Dubreil's article was no longer entirely adequate and it disappeared entirely from the 1996 edition where, *s. v.*, "Emmy Noether," the new article contained only cross-references to other articles in the *Encyclopaedia* in which her name appears. But the 2002 edition rectifies the omission of the *Invariante Variationsprobleme* in the preceding editions with a short article by Bernard Pire who emphasizes the importance of that article and, *s. v.*, "Noether, théorème de," cross-references to "mécanique analytique" and also to the article on Max Noether. However, the electronic version of that encyclopedia that became available in 2006 (version 11) still contained Dubreil's article.

5.3 Analysis of Several Works in Mathematics and Mechanics, 1950–1980

The manner in which conservation laws are treated in the books that we have consulted shows that, in several cases, their authors became aware of Noether's paper when they prepared second or third editions, as though it had not been known to them when they published their first editions.

Gel'fand and Fomin 1961 — The textbook by Gel'fand[4] and Fomin [1961], translated and published in English in 1963, gives a modern presentation of the calculus of variations with applications to physics and mechanics. The authors allow the transformations they consider to depend not only on the independent and dependent variables, but also on the first derivatives of the latter, in effect considering first-order generalized symmetries (p. 177). The book contains a presentation of Noether's first theorem for simple integrals in chapter 4, and in chapter 7 for variational problems in several independent variables. In a note (p. 189), Gel'fand and Fomin state that "the Maxwell equations are actually invariant under a 15-parameter family (group) of transformations" and that there are therefore 5 conservation laws in additon to the 10 derived from the invariance of the equations under the Lorentz group, for which result they cite Bessel-Hagen [1921].

Concerning Noether's second theorem, in the English edition, in a section entitled "Noether's Theorem," one finds, "Let us suppose that the functional $J[u]$ is

---

[3] In his lecture of 22 May 1985 at the Institut Henri Poincaré, Dubreil [1986] mentioned Noether's two articles of 1918 on differential invariants only in passing, and he declared with condescension, "All this is surely very good work" ("Tout ceci est certainement du très bon travail") and then went on to add that that period corresponds to the years of gestation of her more original work.

[4] Gel'fand was one of the great mathematicians of the twentieth century. He died 5 October 2009, at the age of 96, in Highland Park, NJ, close to Rutgers University, where he had come as a professor in 1990. He was born in Krasnye Okny, in Ukraine, and was a professor at Moscow University from 1941 to 1990. When he emigrated to the United States, he first held visiting positions at Harvard University and the Massachusetts Institute of Technology. His work is enormous and ranges from functional analysis and representation theory to biology.

invariant under a family of transformations depending on r arbitrary functions instead of r parameters. Then, according to another theorem of Noether (which will not be proved here), there are r identities connecting the left-hand sides of the Euler equations corresponding to $J[u]$,"[5] after which Gel'fand and Fomin briefly treat the example of homogeneous Lagrangians. In the Russian edition one also finds, at the end of this section,[6] that "we will not tarry here over the proof of Noether's theorem for transformation groups that depend on arbitrary functions, nor shall we write in their general form those relations among Euler's equations of which Noether's second theorem affirms the existence." Though the translator has improved the overall presentation and the notation of the original Russian text in many ways, he also suppressed these lines and replaced them with the incidental clause, "which will not be proved here," which in effect, diminishes the importance imparted to the result of the second theorem. In addition, while the original Russian version had no references, the English edition contains a short bibliography which, despite the discussion of Noether's first theorem, does not include a reference to the *Invariante Variationsprobleme*. Despite that bibliographic omission, Gel'fand and Fomin are sufficiently precise here to suggest that at least one of them had read and understood that article already in the late 1950s, when they were drafting their book.

A further reflection of Noether's presence can be found in the fact remarked *supra* that, after 1975, Noether's expression, "the formal calculus of variations" that we have identified in both [1918c] and [1923], recurs in Gel'fand's articles, first in two papers in collaboration with Leonid Dikiĭ (Dickey) [1975] [1976], then in those in collaboration with Yuri I. Manin and Mikhail A. Shubin [1976] and with Irene Ya. Dorfman [1979]. Alexander Vinogradov recalls on the contrary that Gel'fand's use of the expression "the formal calculus of variations" is tied to his use of the expression "formal geometry," the subject of his lecture before the International Congress of Mathematicians held in Nice in 1970, and that it was independent of Noether's use of that expression.[7]

Funk 1962 — Paul Funk (1886–1969), who had been a student of Hilbert and had defended his doctoral thesis in 1911 in Göttingen, was a professor emeritus at the University of Vienna when he published a treatise, "The Calculus of Variations and its Application in Physics and Technology" [1962], whose section VI.3, "The Theorems of E. Noether" ("Die Sätze von E. Noether"), contains two parts, the first dealing with her first theorem and the second with the second theorem.[8] In

---

[5] Chapter 7, section 37, remark 4, p. 179. In another remark on p. 189, Gel'fand and Fomin apply the preceding remark to the case of the invariance of the Lagrangian density of the electromagnetic field under gauge transformations, "the invariance under gauge transformations (which depend on one arbitrary function) implies the existence of a relation between the left-hand sides of the corresponding Euler equations."

[6] Section 33 of chapter 7.

[7] We thank Professor Vinogradov for sharing with us in an interview on 18 October 2000 his recollections of mathematical research in the circle around Gel'fand in the 1970s. His recollections correspond to what Gel'fand assured us in May 2008, in his home in Highland Park, NJ, that his articles with Dickey were independent of Noether's article.

[8] The theorems are stated on pp. 441 and 446, respectively.

pp. 448–449 he gave some historical remarks, mentioning Hilbert and Klein and the theory of relativity, then, p. 451, he mentions Einstein. The bibliographic references for this part of his book include Noether [1918c], Lie's treatise on transformation groups [1893] and several volumes of his collected works, and Bessel-Hagen [1921]. In addition, he cites articles by Johann Radon dating back to 1928 and 1938, but these do not deal with problems of invariance, so they do not refer to Noether. This textbook in German offers a reasonably complete exposition of Noether's results[9] and situates them, as is relevant, in the context of the development of the general theory of relativity, but it seems to have had very limited influence among mathematicians and physicists, perhaps because it was never translated into English.

Souriau 1964, 1966, 1970, 1974 — In a note in the *Comptes rendus* of the Paris Academy of Sciences [1957], Jean-Marie Souriau extended the concept of an energy-momentum tensor and the differential identities that are associated with it to various phenomena, but without providing a reference to Noether or to any other source. In 1964 he rectified that omission:

> Let us assume that a partial system is governed by a variational principle, and that there exists an infinitesimal transformation which leaves the system invariant. The methods of Emmy Noether [with a footnote referring to her article] then permit defining a certain quantity [...] and proving that the numerical value of that quantity is constant (one says that this quantity is *conserved*).[10]

On the next page he stated "the Noetherian character" ("le caractère noethérien") of the quantities, energy, linear momentum, and angular momentum, "which are in one-to-one correspondence with the invariances of the system" ("qui correspondent biunivoquement aux invariances du système"). This is, therefore, a clear reference to Noether's first theorem and a recognition of the importance of her work. Two years later, before treating geometric quantization, Souriau wrote, "One knows that Noether's theorem shows that [energy] is conserved in any variational problem, on the condition that the Lagrangian be invariant under translation of the variable t."[11]

However, in his fundamental book, "Structure des systèmes dynamiques" [1970], in a section dealing with the calculus of variations in which he treated the "transformations of a variational problem" ("transformations d'un problème variationnel"), Souriau stated a form of Noether's theorem relative to classical mechanics.[12] His

---

[9] The fact that this comprehensive treatise contains "the formulation of the conservation theorems of E. Noether under a group of transformations" ("un exposé des théorèmes d'E. Noether de conservation par un groupe de transformations") was mentioned by René Deheuvels in his review of this book for volume 27 of the *Mathematical Reviews*.

[10] "Supposons qu'un système partiel soit régi par un principe variationnel, et qu'il existe une transformation infinitésimale qui laisse le système invariant. Les méthodes d'Emmy Noether [in a footnote, the reference to her article] permettent alors de définir une certaine quantité [...] et de démontrer que la valeur numérique de cette grandeur est constante (on dit que la grandeur est *conservative*)," Souriau [1964], p. 318.

[11] "On sait que le théorème de Noether montre que [l'énergie] est conservative dans tout problème variationnel–pourvu que le lagrangien soit invariant par translation de la variable t," Souriau [1966], p. 375.

[12] Souriau [1970], p. 72; English translation, p.68.

formulation both modernized and restricted the result of Noether's first theorem. His treatment was modern because he worked in the framework of manifolds, just as Andrzej Trautman had done, using the language of jets, in his 1967 article which will be analyzed shortly.[13] Yet, because he considered Lagrangians defined on the tangent bundle of the configuration manifold, he was in fact considering only the case of a single independent variable, time, and of a first-order, autonomous, i.e., not depending explicitly on time, Lagrangian. In addition, he considered only invariance under diffeomorphisms which are the prolongations of diffeomorphisms of the configuration manifold, and therefore he did not treat infinitesimal invariance under generalized vector fields.

On the other hand, we owe to both Bertram Kostant and to Souriau,[14] working independently, the introduction of the concept of the moment of a dynamical group defined in the case of a Hamiltonian group action. The map thus defined from the phase space to the dual of the Lie algebra of infinitesimal symmetries of the system is called the "moment map" (after the French, "application moment") or the "momentum map." It is in this context that Souriau generalized Noether's theorem in a Hamiltonian form as follows: the moment of a Hamiltonian action is conserved. Four years later he called this result the "symplectic Noether theorem" ("théorème de Noether symplectique").[15]

Thus, under the influence of Souriau's book, the practice arose of calling "the Noether theorem"[16] the following result, whose proof is an immediate consequence of the skew-symmetry of the Poisson bracket: for a dynamical system with Hamiltonian H, the Hamiltonian vector field associated with any conserved quantity K generates a one-parameter group of symmetries of the system. This result could possibly be called "the Hamiltonian Noether thoerem." The practice of calling it "the Noether theorem" is regrettable because, in fact, the Hamiltonian point of view appears nowhere in Noether's work, and it is therefore inappropriate to give her name to this important, yet easily proved result.

Lanczos 1966 — The first two editions (1949 and 1962) of Lanczos's treatise on *The Variational Principles of Mechanics* [1949] do not take into account Noether's results, and contain no reference to her article. However, the third and fourth editions (1966 and 1970) include an Appendix II entitled "Noether's invariant variational problems" which, of course, cites her article. In the preface to the third edition (dated 1965, reproduced in part on p. xii of the 1970 edition), Lanczos writes, "The present edition differs form the previous one by the addition of a section on Noether's invariant variational problems. The original paper of Noether is not easy reading. In fact, however, the discussion [of her first theorem] can be subordinated to the well-known theory of 'ignorable variables,' and this is the method followed in the present

---

[13] See, *infra*, p. 110.

[14] Souriau [1970], pp. 105–107, Kostant [1970], pp. 176 and 187. Also see the article of the same period by Stephen Smale [1970]. For a brief history of the concept of moment map, see Marsden and Ratiu [1999], p. 369, and for modern developments see Weinstein [2005].

[15] Souriau [1974], p. 357.

[16] See, for example, Marle [1983], p. 161, Libermann and Marle [1987], p. 197, or Cushman and Bates [1997], p. 406.

exposition. In this approach the parameters of Noether's transformation appear as added action variables of the variational problem, for which the Euler–Lagrange equations can be found." Lanczos continues in this preface (but not in the 1970 edition) to argue that, because of its "important applications" in "the field theories of modern physics," the generalization to the case of partial differential operators, i.e., to the case of several independent variables, should be dealt with. In his Appendix II of the third edition he repeats this argument (p. 357) and treats (p. 361) the case of a (first-order) Lagrangian with time as the single independent variable, when that Lagrangian does not depend explicitly on time, or is invariant under translations, or rotations, by the method of "kinosthenic" variables described in his chapter V. These had been considered by Routh [1877][17] and Helmholtz [1884], and were called "ignorable or cyclic" by Whittaker [1904] (2nd ed., 1917, p. 104), while "kinosthenic or speed coordinates" were the terms introduced by J. J. Thomson [1888].[18]

The fourth edition of Lanczos's book (1970) contains, in addition to Appendix II (pp. 401–405), a section 20 of chapter XI, pp. 384–386, on "Noether's principle," where he determines the conservation laws associated with the invariance of the Lagrangian under multiplication by a phase factor for Maxwell's equations and for the Schrödinger equation.

Rund 1966 — In his book on Hamilton–Jacobi theory [1966], Hanno Rund gives an account of "Noether's theorem," first in the case of a single independent variable, then for "multiple integral problems."[19] He proves her first theorem for Lagrangians invariant up to an "independent integral," of which a divergence is a particular case, and he states as a corollary that an invariance of a Lagrangian implies a conservation law for the associated Euler–Lagrange equations. In a remark (p. 297) he points out that the form of Noether's theorem that he obtained as his corollary can be "generalized in a different direction" by considering transformations of a more general type just as Gel'fand and Fomin [1961] had done for the case of a single independent variable, and he adds, "For most applications of the theorem this type of generalization is not necessary and will therefore be ignored here." Thus Rund hinted at a particular case of the generalized symmetries with which Noether had dealt, but he did not consider it worthwhile to develop this remark.

Some of this material is also in the book on variational principles by David Lovelock and Rund [1975], where they discuss "the famous theorems of Noether." They cite Noether's article in the chapter on variational problems invariant under some "*r*-parameter transformation group" (chapter 6), where they claim to use a "simple and direct approach" instead of "the original derivation of Noether, which is fairly complicated and depends on some deep and difficult theorems in the calculus of variations." When treating "Invariant variational principles and physical field theories" (chapter 8), they derive "invariance identities" in order to determine which

---

[17] Edward J. Routh (1831–1907), in his book on the stability of motion [1877], chapter 4, paragraph 20, had treated the case where "the Lagrangian function $T - V$ is not a function of some of the coefficients as θ, ϕ, &. though it is a function of their differential coefficients θ', ϕ', &."

[18] Joseph John Thomson (1856–1940) received the Nobel Prize in physics in 1906.

[19] P. 73 and chapter 4, section 6, "The theorem of Noether," p. 293.

Lagrangians satisfy a given invariance, but do not mention Noether's second theorem when treating the case of the Einstein–Maxwell field equations. It may be that by "the famous theorems of Noether," they only mean the first theorem in the case of either a single or a multiple integral.

Trautman 1967 — Beginning in the 1950s, in notes published in the *Bulletin de l'Académie Polonaise des Sciences* [1956] [1957], Trautman studied conservation laws in general relativity, and then, in 1962, published a long article on this subject that we shall analyze below.[20] Here, we consider his "Noether equations and conservation laws" [1967] where he was the first to present even a part of Noether's article in a modern mathematical language, the language of manifolds, i.e., without introducing local coordinates. In this article and in a second that he would publish five years later [1972], Trautman also used, for the first time in the literature dealing with Noether's theorems, the theory of fiber bundles and, in particular, the jet bundles that had been defined by Ehresmann in the 1950s.[21] However, Trautman's 1967 article modernized only a small part of Noether's theory because it only considered the case of first-order Lagrangians that are invariant with respect to classical symmetries. These two articles were followed by a note written with Demeter Krupka [1974], where the authors study those Lagrangians on fiber bundles that are invariant under all transformations of the base manifold, and where, in the concluding section, they refer to Noether's article [1918c] and urge the reader to compare their approach with hers. After these pioneering works, numerous other authors published articles of increasing generality, setting Noether's results in an invariant, geometric language.

Edelen 1969 — D. G. B. Edelen's book "Nonlocal Variations and Local Invariance of Fields" [1969] has no references and a very idiosyncratic terminology. His aim is mainly to reveal the geometric nature of "the functions occurring as arguments of the functionals under consideration" when they "are changed by a coordinate transformation or a point transformation acting on the manifold of independent variables." However, his chapter on "Invariance considerations" (chapter 4) opens with the sentence, "Repeated use is made of Emmy Noether theorems in many contexts," and there he proceeds to discuss several types of invariance that actually correspond to the two cases of Noether's theorems.

Smale 1970 — In 1970, Stephen Smale published a two-part article entitled "Topology and mechanics" [1970] which proved to be very influential. In the section entitled "Symmetry in mechanics" he first asserted that "Noether is largely responsible for clarifying the relationship between symmetry and integrals," and then he treated mechanical systems with symmetries using the tangent bundle to the configuration manifold, and he introduced the concept of "angular momentum." His approach is intermediate between a purely Lagrangian approach in terms of the tangent bundle, and the Hamiltonian formalism that uses the cotangent bundle.

---

[20] See, *infra*, Chap. 6, p. 126.

[21] See Ehresmann [1951]. Fiber bundles were introduced by Ehresmann (1905–1979) and by his very gifted student, Jacques Feldbau (1914–1945), even earlier, just before 1940, and simultaneously by Norman Steenrod, following previous work of Hassler Whitney and Eduard Stiefel.

Śniatycki 1970 — In his article [1970], Jedrzej Śniatycki contributed to an invariant geometric formulation of the calculus of variations using jet bundles, following the work of De Donder [1935], Weyl [1935b], Théodore H. J. Lepage [1936] and Trautman [1967]. He, too, limited himself to first-order Lagrangians but reformulated the results of Noether's first theorem for generalized symmetries of order k, asserting that he thus obtained "not only the results of Noether's first theorem, but also the conservation laws for which the generators of the symmetry transformations depend on higher-order derivatives." He attributed the introduction of generalized symmetries to Heinz Steudel [1966] and to the physicist Arthur Komar, in an article on the problem of quantization in general relativity [1967], and he also cited Jan Komorowski [1968]. His result is in fact a useful translation of Noether's 1918 result into invariant geometric form expressed in the language of manifolds and jet bundles.

Krupka 1971, 1973 — Krupka's first publication [1971], which proceeds from Gel'fand and Fomin's "Calculus of Variations" [1967], from Trautman's intrinsic treatment of Noether's theorem [1967] and from Ivan Kolář's work on the theory of jet bundles, deals with several aspects of the Lagrangian formalism in the framework of fiber bundles. Noether is credited with a "Noether equation" in a form in fact due to Trautman, concerning the transformation of a Lagrangian under a local diffeomorphism of the first jet bundle of a fiber bundle, but she is not cited in the short section on conservation laws. However, in his monograph [1974], Krupka gives a more comprehensive account of the geometric theory of variational problems, and concludes with a section on invariant variational problems where Trautman's approach is significantly generalized, while both Noether's 1918 article and Śniatycki's 1970 article are listed in the references. But Krupka's first paper was not cited by Goldschmidt and Sternberg [1973], and neither of Krupka's papers seems to have been known to García[22] in 1974.

Goldschmidt and Sternberg 1973 — In a path-breaking article, Hubert Goldschmidt and Shlomo Sternberg [1973] gave "an exposition of the geometry of the calculus of variations in several variables" with "the consistent use of fibered manifods and the affine structure of jet bundles, " and they announced that their article includes a discussion of "Noether's theorem." In the short list of references which appear in the introduction, the earliest are Cartan's book of 1922, Carathéodory's and De Donder's books of 1935, and Weyl's [1935b], but not Noether's original article, and the list of later references they cite includes Hill's 1951 article. They proposed a general geometric formulation of the Lagrangian theory when the Lagrangian is defined on the first jet bundle of a fibered manifold, i.e., when it depends on sections of a fibered manifold and on their first derivatives. In this case they were in fact able to formulate Noether's first theorem in an intrinsic manner. They then defined the Hamilton–Cartan form and proved "a more general version of Noether's theorem," in which they succeeded in giving a coordinate-invariant formulation of the conservation law corresponding to an invariance property up to divergence. This

---

[22] See *infra*.

article was thus an important step in the development of the geometric formulation of Noether's first theorem and of Bessel-Hagen's generalization.

García 1974 — In his contribution [1974] to the international Rome conference of 1973 on symplectic geometry, Pedro García adopted a point of view that emphasizes "the conceptual identity between the modern 'current algebras' and the old 'Noether theory' on invariant variational problems." However, he did not give an explicit reference to Noether's article, or references to work on the subject that are anterior to 1960, but he did include references to articles by I. E. Segal, Souriau, García and Antonio Pérez-Rendón, and Robert Hermann, all published between 1968 and 1971 and thus contemporary to Śniatycki's treatment of the Noether theorems. García's article contains a detailed study, in invariant geometric terms, of the map which to an infinitesimal symmetry, X, associates a "Noether invariant," seen as the differential form obtained as the interior product of the generalized Poincaré–Cartan form by the vector field X. García defined trivial conservation laws, and demonstrated the existence of a Poisson bracket on the space of conservation laws modulo the trivial conservation laws.

The articles by Trautman [1967], Śniatycki [1970], Goldschmidt and Sternberg [1973], and García [1974] in the area of differential geometry were followed by articles in the same area by Krupka [1975] and Kolář [1984] among so many others that they cannot all be cited here.[23]

Arnold 1974 — In the chapter entitled "Lagrangian mechanics on manifolds" of his book on the mathematical methods of mechanics [1974], Vladimir Arnold[24] devoted a section to "E. Noether's theorem." The hypothesis of the theorem as he stated it has the same generality as in Souriau's formulation [1970]. The case where the Lagrangian depends explicitly on time was treated as an exercise. Just as Souriau's book, Arnold's contains very few references and none to Noether. The German translation of 1988 of the second Russian edition (1979) presents (p. 288), under the description "generalization of Noether's theorem," the Hamiltonian version of Noether's first theorem that had already been stated by Souriau, but this subtitle is not used in the Russian original.

Marsden 1974, 1978 — Another line of references, which derives from Souriau's and Smale's treatment of the first Noether theorem, leads to the work of Jerrold E. Marsden [1974], to the book by Paul Chernoff and Marsden [1974], and to the second edition of *Foundations of Mechanics* by Ralph Abraham and Marsden (1978).

In his monograph on infinite-dimensional Hamiltonian systems [1974], Marsden referred to work on this subject which he had published already in 1968, to Souriau [1970] and to Smale [1970], and he wrote that he mainly followed Souriau [1970], so that "Noether's theorem" (pp. 145–146) is actually the Hamiltonian version of

---

[23] Numerous additional references may be found in Kosmann-Schwarzbach [1985] and [1987].

[24] Born in Moscow in 1937, Arnold died in Paris on 3 June 2010. He produced deep work in such diverse fields as differential equations, Hamiltonian dynamics, hydrodynamics, mechanics, differential geometry, topology and singularity theory.

the conservation theorem, which we have called, *supra*, the "Hamiltonian Noether theorem."

In their volume in the series "Lecture Notes in Mathematics" [1974], Chernoff and Marsden elaborate a precise analytic study of the moment map in the sense of Souriau, in the infinite-dimensional case, taking into account the domains of definition of the operators, more specifically, in the case where the Hamiltonian system being considered is defined on an infinite-dimensional Banach manifold equipped with a weak symplectic structure. The associated Hamiltonian systems are partial differential equations, such as the Schrödinger equation. The last section of their book (p. 149) is entitled "Noether's theorem and local conservation laws," and it begins with the remark, "Judging by the number of papers devoted to it, Noether's [first] theorem must be one of the most popular propositions of all time." Chernoff and Marsden offer an invariant version of the first theorem on a vector bundle, where they deal with the case of a Lagrangian of order one and of classical symmetries, and they state a version of Noether's theorem in the form of a local conservation law, $\frac{\partial \mathscr{I}}{\partial t} + \mathrm{Div}\, T = 0$. They conclude their book with the remark, "Of course a great deal more can be said about Noether's [first] theorem and the rich geometrical ideas which underlie it." Their references are García [1974], Hermann [1970c][25] and Trautman [1967].

While the first edition of Abraham and Marsden's book [1967] contained only a discussion of conservation theorems for Hamiltonian systems and thus no mention of Noether's theorem, the second edition (1978) contained a geometric description close to that of Smale. The authors first stated Noether's theorem for a Lagrangian defined on a tangent bundle (p. 285). In this case, the Legendre transform, $\mathscr{F}L$, is identified with the fiber derivative of L. The infinitesimal transformation that leaves the Lagrangian invariant is assumed to be vertical, i.e., it does not act on the time variable, which is in fact the independent variable. Then, in section 5.5, p. 479, the authors stated a version of Noether's first theorem that is applicable to special relativity and is similar to the one that had appeared in Chernoff and Marsden [1974].

Logan 1977 — John David Logan was clearly well informed about Noether's work. He wrote his doctoral thesis in 1970 on *Noether's Theorems and the Calculus of Variations* under the direction of Stefan Drobot, a student of Hugo Steinhaus, himself a student of Hilbert, who defended his doctorate in 1911 in Göttingen. Soon after his thesis, Logan published several articles on invariant variational problems, including a version of Noether's theorem for discrete systems [1973], and an article giving an alternative proof of Noether's second theorem [1974].

In the preface to his book *Invariant Variational Principles*" [1977], Logan acknowledges Noether's "monumental" article, and asserts that "the Noether theorem has become one of the basic building blocks of modern field theories." Then he cites in his references the original text of the *Invariante Variationsprobleme*, M. A. Tavel's English translation of that article (Noether [1971]), as well as Tavel's historical analysis [1971a]. Logan offers a detailed analysis of the case of

---

[25] See, *infra*, Chap. 7, p. 140.

multiple integrals with an arbitrary number of independent variables for a first-order Lagrangian invariant under classical symmetries; then he treats second-order Lagrangians in the case of a single independent variable (a simple integral), applies the result to obtain several conservation laws for the Korteweg–de Vries equation which, written in potential form, derives from a second-order Lagrangian. Finally he gives an idea of Noether's second theorem "in a special case," gives an application of the second theorem to electromagnetism, and discusses Noether's example, the Weierstrass condition for homogeneous Lagrangians. Logan refers not only to Noether but also to the English translation (1953) of Courant and Hilbert's manual [1924], to Funk [1962], to the translation of Gel'fand and Fomin [1961], to Rund [1966], to Sagan [1969] and to Hill [1951], as well as to the modern treatments by Trautman [1967], Komorowski [1968], García [1968], Edelen [1969], and Chernoff and Marsden [1974]. He also cites Drobot and Rybarski [1958], and remarks that the application to the n-body problem of the search for conservation laws associated to symmetries was first accomplished by Bessel-Hagen in 1921.[26] In this connection, Logan cites Whittaker's *A Treatise on the Analytical Dynamics of Particles and Rigid Bodies* [1917], where one indeed finds, in chapter 3, a discussion of conservation laws.[27] Logan recognizes Noether's originality, "Invariant higher-order problems were first considered by Noether in her original paper," and immediately adds, referring to Dan Anderson's article [1973], that "Anderson puts Noether's work in more modern form." But this statement is misleading because Anderson's paper is only one of many publications that claim to generalize Noether's results, while they are just rediscoveries of what Noether's article already contained. (See, *infra*, p. 121.)

Ovsjannikov, Ibragimov — In section 30, the last, of his book [1978], translated into English in 1982, Lev Vasil'evich Ovsjannikov deals with conservation laws, offers a somewhat modernized version of Noether's first theorem, and refers to the Russian translation of Noether's article in Polak [1959]. Then he states a form of the converse of her theorem, and he treats an example drawn from the theory of gas dynamics. He refers to several articles by his student, Nail H. Ibragimov, published since 1969. The latter, in his book [1983] based on lectures delivered since 1972 and on articles published in Russian and English in 1969, 1976, 1977 and 1979, presents a complete version of Noether's first theorem, but with an error in the general formula for what he calls the "Noether operators," i.e., the operators associated to a symmetry which, when applied to a Lagrangian, yield the several components of the conservation laws. His Lagrangians are of arbitrary order, and his infinitesimal symmetries are the generalized symmetries as Noether had introduced them in her point of departure, and he calls them Lie–Bäcklund operators.[28] He studies the action of the symmetries

---

[26] See, *supra*, Chap. 4, p. 92, note 4, where it is shown that Drobot and Rybarski had already recognized not only the role of Bessel-Hagen, but also that of Noether in the introduction of symmetries up to divergence.

[27] While the first and second editions of Whittaker's book predate the publication of Noether's results, there were many editions of his treatise published after 1918 that could have referred to her work, but did not.

[28] See, *infra*, Chap. 7, p. 134, for Lie–Bäcklund and generalized symmetries.

on the conservation laws (actions of the adjoint algebra), just as Noether had done in the fifth section of her article, mentions in a footnote that he will not consider the case of infinite groups depending on arbitrary functions, and refers to Noether's treatment of this case.[29] Then he treats the case of evolution equations and finally he offers applications of the theorem.

Kijowski and Tulczyjew 1979 — In their book *A Symplectic Framework for Field Theories* [1979], Jerzy Kijowski and Włodzimierz Tulczyjew offer "a systematic procedure for deriving variational formulations of physical theories" (p. 1) and develop a new physical theory for both particle dynamics and field theory based on the concept of Lagrangian submanifolds of symplectic manifolds. This theory is the result of a collaboration between a mathematical physicist, Kijowski, and a mathematician, Tulczyjew. They define an energy-momentum density in sections 19 and 20, then in their section 21, on conservation laws, show that it is conserved under symmetries, and they conclude that "conservation laws can be considered a part of the Hamilton–Jacobi theory." They then add, "Usually these laws are derived within the framework of the Lagrangian formulation of field theory using Noether's theorems," and they explicitly cite Noether and Bessel-Hagen in their vast list of references in mathematics and physics. However, they do not develop the Lagrangian theory and, for a "modern formulation of Noether's theorems," they refer (p. 157) to the work of Trautman [1962] [1967] and García [1974], and to Robert Hermann's book on Lie algebras and quantum mechanics [1970b]. These references were indeed the first steps toward a differential-geometric formulation of Noether's first theorem that culminated in the work of Vinogradov and Tsujishita. However, they contain no hint of the existence of the second theorem.

Goldstein 1950, 1980, 2002 — The first edition of Herbert Goldstein's textbook of mechanics [1950] has often been reprinted and was translated into French in 1964. It contains a discussion of the several conservation theorems in mechanics, but contains no statement bearing the name of Noether's theorem.[30] But Goldstein's second edition (1980) opens with a long preface in which he explains the logic of the changes that he introduced in that edition, and to whom the book is addressed, and says that he refrained from introducing the most modern mathematics, the tools of differential geometry and topology, even though they are used in modern theoretical physics, keeping the mathematical demands on his readers to a minimum. However, he radically revised his point of view concerning conservation laws and included a section on "Noether's theorem." Goldstein emphasizes in his preface (p. viii) that the section in question (section 12.7) is new, and he adds in a bibliographical note (p. 597) that he has followed Boyer [1966], and that "Noether's theorem

---

[29] "When the elementary action is invariant under such a group, a dependence between the left-hand sides of [the Euler–Lagrange equations] is observed," p. 320 of the 1985 English translation of Ibragimov [1983].

[30] See Goldstein [1950], pp. 47, 220 and 261. Similarly, another, more specialized book, on mechanics oriented toward applications in aeronautics, by Sanford W. Groesberg [1968], which appeared between the first and second editions of Goldstein's book, contained a Lagrangian theory and the derivation of various conservation laws, but without a general theory and thus contained no reference to Noether.

is discussed, but not by name" by Asim O. Barut [1964].[31] In this second edition Noether's reputation has caught up with her, and Goldstein echoes Chernoff and Marsden's and Logan's praise when he describes her as "one of the leading mathematicians of this century,"[32] and proceeds to a thorough discussion of "Noether's theorem," i.e., Noether's first theorem. In fact, however, he restricts his discussion to the symmetries of the Lagrangian which are still only the classical symmetries, and to first-order Lagrangians, but he adds (p. 589) that this method may be extended to symmetries up to divergence, and that the version of Noether's theorem that he presents is not the most general one possible. The footnote in section 2.6 of the original edition concerning cyclic or ignorable variables, citing books by A. G. Webster, by W. E. Byerly, and by Ames and Murnaghan, is retained in the 1980 edition even though their results are subsumed in Noether's, and he adds a reference to Lanczos, probably to the 1966 edition of Lanczos [1949]. Thus one may speculate that it was Lanczos who introduced Goldstein to Noether's first theorem, in fact to a restricted form of that theorem. Goldstein may of course have read about the Noether theorem or even theorems elsewhere, or heard about them from colleagues, but those sources have left no trace in his book. He does not seem to know Noether's paper because, when he discusses soliton equations (Korteweg–de Vries and sine–Gordon), he observes that, aside from the translation symmetries for Korteweg–de Vries and the Lorentz transformations for sine–Gordon, no other symmetries for these two equations are known even though an infinite number of conservation laws are known for them, "so that the last word has probably not yet been said on the relation between conserved quantities and the nature of the field" (p. 595). This implies that he did not know about Noether's generalized symmetries, nor that they had been rediscovered by Anderson, Kumei and Wulfman [1972], had been studied by Anderson and Ibragimov [1979] under the name of Lie–Bäcklund symmetries, and had been used in their algebraic form by Manin and his school,[33] and in their geometric form as tangent vectors to jet bundles by Vinogradov and his school since the 1970s.[34] Ignorance of some recent and, to a degree, some classical results is inevitable, but the consequence was that here Goldstein left the impression that a major problem that had been solved by Noether remained to be solved by future generations of mathematicians.

In a third edition, published under the names of Goldstein, Charles Pooles and John Safko [2002], the passages upon which we just commented remained essentially unchanged, but the footnote of section 2.6 with its references was omitted, while the footnote at the beginning of section 12.6 that contained praise of Noether has, strangely, also disappeared.

---

[31] Indeed Barut in his book [1964] describes results concerning conservation laws and, although he does not supply any precise attributions, Noether's article does appear in a list of references (p. 130).

[32] Goldstein [1950], 2nd ed., 1980, section 12.6, p. 588, note.

[33] See, *infra*, Chap. 6, p. 139.

[34] See, *infra*, Chap. 6, p. 143.

We shall cite in its entirety the concluding paragraph of the second edition (1980, p. 596) which remained unchanged in the third, becuse of its relevance to the thesis of this book:

> Thus, the theorems on the conservation both of Jacobi's integral and of the generalized momentum conjugate to a cyclic coordinate are subsumed under Noether's theorem as stated in Eq. (12-164). The connection between symmetry properties of a mechanical system and conserved quantities has run as a thread throughout formulations of mechanics as presented [in this book]. Having come full circle, as it were, and rederived by sophisticated techniques symmetry theorems found in the first chapters, it seems an appropriate point at which to end our discussions.

We have thus shown that the period between 1950 and 1980 was one of a gradual development in which the work of Noether on the calculus of variations was recognized and cited. The re-issue of Goldstein's book in 1980 is an illustration of that process.

5.4 Analysis of Several Works in Physics, 1950–1980

Here, too, we must note many negative results in a search for references to Noether's theorems. For example, in his classical treatise on quantum mechanics, Leonard I. Schiff [1949] recalls no results from classical mechanics, but the note on p. 133, concerning variational problems, urges consultation of Whittaker (3rd ed., 1927), the manual of H. C. Corben and Philip Stehle (1st ed., 1950, 2nd ed., 1960) or the book by Goldstein [1950] which we just analyzed. In another classical manual, Lev D. Landau and Evgeni M. Lifchitz's *Field Theory* [1948] (first English translation, 1951), the authors indeed discuss the energy-momentum tensor but do not mention the theorems of Noether or their connection with relativity.

In his "Relativistic Theories of Gravitation and Electro-Magnetism" [1955], André Lichnerowicz (1915–1998) discussed the identities satisfied by generally invariant Lagrangian equations with the following comment:

> This process leads to equations that are invariant under all changes of admissible coordinates, and the left-hand sides automatically satisfy 4 conservation identities which we shall derive by a method the principle of which goes back to Hermann Weyl,[35]

thus attributing the discovery of these identities to Weyl, while Noether's contemporaneous derivation was in fact more general.

Richard Feynman, in several sections of volume 1 of his course *Lectures on Physics* [1963],[36] discussed the conservation laws of classical mechanics, but always with an absolute minimum of mathematical formalism. Then, in Section 52-3, entitled "Symmetry and conservation laws," Feynman wrote,

---

[35] "Ce procédé conduit à des équations invariantes par les changements de coordonnées admissibles et les premiers membres satisfont automatiquement à 4 identités que nous allons former par une méthode dont le principe remonte à Hermann Weyl," Lichnerowicz [1955], p. 270.

[36] Feynman designed this two-year course for undergraduates at the California Institute of Technology who would not all be physics majors.

> The symmetries of the physical laws are very interesting at this level [classical mechanics], but they turn out, in the end, to be even more interesting and exciting when we come to quantum mechanics. For a reason which we cannot make clear at the level of the present discussion—a fact that most physicists still find somewhat staggering, a most profound and beautiful thing is that, in quantum mechanics, *for each of the rules of symmetry there is a corresponding conservation law*; there is a definite connection between the laws of conservation and the symmetries of physical laws.

He then gave the examples of the conservation of linear momentum, energy, angular momentum and electrical charge. Finally, in chapter 17 of volume 3, he treated the correspondence between symmetries of the Hamiltonian and constants of the motion defined by the Hamiltonian, but he did not introduce the Lagrangian formalism, so it is not surprising that Noether's name does not appear there either.

As an example of Hill's article serving as a screen between the physicists and Noether, we shall quote from an article by M. A. Melvin [1960]. In his part C, "Survey of theoretical background and connection between symmetry principles and constants of motion," he simply writes,

> The association between invariance and conservation points of view has its prototype in classical Hamiltonian dynamics; this is thoroughly discussed in Hill (1949) [*sic* for 1951],

and then goes on to discuss the situation in quantum mechanics.

Iwanenko and Sokolov 1953 — In their study of classical electrodynamics in *Klassische Feldtheorie* [1953], Dmitri Iwanenko and Arsenyi Sokolov use Noether's theorem, i.e., her first theorem, list the reference to Noether's article explicitly, and call attention to its use by Bessel-Hagen [1921] to derive the conservation laws of Maxwell's theory of electrodynamics in classical physics, and then by Markow [1936] to describe Dirac's theory of the electron in quantum mechanics.

Rzewuski 1953, 1958 — After publishing an article on conservation laws [1953] which contains a reference to Noether, Jan Rzewuski published *Field Theory, I: Classical Theory* [1958] in which he speaks about "the most important gauge transformation groups and the translations and rotations of space-time." For "the general case of arbitrary continuous groups," he refers to Noether with a reference in a footnote, but he omits her name from his index.

Winogradzki 1956 — In a short article in French [1956], Judith Winogradzki, who was then working in the Institut Henri Poincaré in Paris,[37] formulated Noether's first theorem in a form adapted to the invariances of special relativity, and then studied its consequences for the energy-momentum tensor. She cited Noether and Hill, using the latter's simplified formulation, and also an article by Paul Roman [1955] anterior to the book to be discussed *infra*. Her article would be cited by Barut [1964].

Bogolyubov and Chirkov 1957, 1959 — Nikolai N. Bogolyubov and Dmitrii V. Chirkov, in their treatise on quantum field theory [1957] that was translated into English in 1959 and then into French in 1960, entitled their section 2.5 "Noether's

---

[37] She was subsequently appointed to a chair of theoretical physics at the University of Rouen, and died in 2006.

Theorems," and wrote, "To construct the invariants of fields we will use the theorem of Noether" without, however, giving a reference. This suggests that some form of Noether's results was well known in the Soviet Union, though possibly more by reputation than by a text. In their bibliography they cite five articles by Pauli, one by Ovsjannikov of 1956, van der Waerden's *Die gruppentheoretische Methode in der Quantenmechanik* [1932], and Gregor Wentzel's book on quantum field theory [1943], which was translated into English in 1949. The Lagrangians that Bogolyubov and Chirkov use are of the first order, and the symmetries they treat are the classical ones. They give several applications of the formula expressing the components of the conservation law associated with an infinitesimal symmetry.

Roman 1960, 1969 — Paul Roman's book on the theory of elementary particles [1960] includes a chapter of nearly 200 pages entitled "Invariance properties and selection rules," in which he begins his discussion with a proof of the first Noether theorem in the case of first-order Lagrangians, apparently following Hill, but he observes that "the relation between continuous symmetry groups and conservation laws was first observed in all generality by E. Noether," and, in a footnote, he refers to her [1918c].[38] He then studies the case of conserved quantities in quantum field theory. He shows that while to every symmetry of the system there corresponds a quantity which is conserved in the time evolution, this conserved quantity corresponds to the integral over a domain in space of the time component of the Noether current. The rest of the chapter studies the physically important cases of symmetries, continuous or discrete.

In his subsequent book on quantum field theory [1969], Roman's only comment relating to Noether is, "The relation between continuous invariance transformations and the continuity equation (2.36) is often referred to as Noether's theorem" (p. 67, note 10), without giving references to justify his use of the adverb, "often."

Boyer 1967 — Timothy H. Boyer, then at Harvard University, published an article [1966] on the subject of symmetries and conservation laws where, "for the sake of brevity and clarity," he reproduced Hill's presentation [1951] of "Noether's theorem," writing that

> The description of the procedure necessary to obtain the conservation laws to which students are urged to refer in courses on electromagnetism and quantum field theory is that of Hill. That article is certainly a complete and precise exposition of the question.

What follows is actually a list of criticisms of that article, which is held to be too difficult, including notation that is contrary to the general practice of physicists and which may lead to confusion. However, a year later we find in section 4 of Boyer's article [1967] an analysis of Noether's (first) theorem stated in a general form, and much of that article is devoted to a careful discussion of the distinction among the symmetries of the Lagrangian function *L*, symmetries of the Lagrangian

---

[38] Fritz Rohrlich, reviewing Roman's earlier paper [1955] for *Mathematical Reviews*, wrote, "When the equations of motion are derivable from a variational principle (Hamilton's principle), it is also possible to derive the conservation laws from that principle," followed by the reference to Noether [1918c] "summarized by E. L. Hill [1951]."

density $L\,dx$, and symmetries of the action integral. Boyer does not seem to know Bessel-Hagen's article [1921]. Here he introduces the vertical representative of a generalized symmetry but, contrary to what Bluman and Kumei thought,[39] this was not the first observation of the fact that generalized symmetries may be replaced by their vertical representatives, because Weyl had already dealt with them in the case of a space-time metric [1917] [1918], Kneser [1918] had used them in the case where time was the single independent variable, and Noether herself had systematically introduced the vertical variations $\bar{\delta}u_i$.[40]

Anderson 1967 — James L. Anderson, who had been a student of Peter G. Bergmann whose publications will be discussed shortly,[41] declared in the preface of his book [1967] that in chapter 4, which would constitute "the foundation chapter for the whole work," he would describe the ideas of covariance and symmetry, and "the relation between symmetry and conservation laws that derives from Noether's theorem." In the introduction to that chapter he explains that, when the elements of a symmetry group of a theory deriving from a variational principle are characterized by one or several arbitrary functions on space-time, the equations of motion are not independent but satisfy "Bianchi-type identities." He then proves "the important identity of Noether" in the classical case of a first-order Lagrangian, and supplies a reference to Noether.[42] Then he discusses "gauge groups," cites Hilbert [1915] in this connection, and gives some indications regarding the difficulty of solving the Cauchy problem[43] due to the presence of identities that are satisfied in covariant gauge theories.

Itzykson and Zuber 1980 – After having shown that the energy-momentum tensor is conserved for a Lagrangian invariant under translations of space-time in special relativity, Claude Itzykson and Jean-Bernard Zuber [1980] argue (p. 23) that "This result is a typical case of Noether's theorem. The latter states that to any continuous one-parameter set of invariances of the Lagrangian is associated a local conserved current. Integrating the fourth component of this current over three-space [i.e., 3-dimensional space] generates a conserved 'charge.'" They then discuss the role of "internal symmetries" which act only on the field variables, and not on the independent variables. For any one-parameter group of such symmetries of the Lagrangian, there also exists a conserved current. The Hamiltonian point of view consists of separating the time variable from the space variables, and seeing the evolution of the system governed by the variational equations derived from the given Lagrangian as

---

[39] Bluman and Kumei [1989], p. 258.

[40] See, *supra*, Chap. 2, notes 10 and 14.

[41] See, *infra*, p. 124.

[42] The reference is exact except for the page number. Anderson may have inherited his error from DeWitt [1964] (see, *infra*, Chap. 6, p. 131, note 17), who may himself have been relying on Rosenfeld [1930], p. 119. Elsewhere, he attributes to "E. Noether" a 1910 article on the notion of a solid body in relativity theory which was in fact the work of her brother, F. Noether.

[43] The Cauchy problem consists in determining the solution of a system of partial differential equations in space-time satisfying initial data conditions on a 3-dimensional hypersurface. More generally, on an n-dimensional manifold, the initial data are specified on an $(n-1)$-dimensional submanifold.

the evolution in time of a system whose initial conditions are fixed on a space-like hypersurface, and which evolves according to the Hamiltonian associated with the Lagrangian. By making time play a special role, on the one hand, one introduces the Poisson brackets of the functionals of fields, and, on the other, one defines the charges as the integrals over space of the time components of the currents. Without supposing that the currents are conserved, it is shown that the Poisson brackets of the time components of the currents associated with one-parameter groups of symmetries reproduce the structure constants of the Lie algebra of the symmetry group. As was said above, it was the quantum version of this phenomenon which, in the 1960s, led to the theory of current algebras. In chapter 11 of their book, Itzykson and Zuber develop the theory of internal symmetries and current algebras in quantum field theory. One should consult the very rich book edited by Manuel Doncel, Armin Hermann, Louis Michel and Abraham Pais [1987] for first-hand accounts of these discoveries, distant descendants of those that Noether had made between 1915 and 1918.

5.5 The Rediscoveries as Generalizations of "Noether's Theorem"

Zaycoff's draft (see Chap. 1, p. 51) was the first of a long series of "rediscoveries as generalizations" of Noether's and Bessel-Hagen's results that would be written in the second half of the twentieth century. Unfortunately, none of those that were published underwent the kind of scrutiny that Noether performed on Zaycoff's paper! All the articles in the period 1950–1975 that claim to generalize "Noether's theorem," always the first, are in fact less general than the theorem which Noether stated and proved, and they are so numerous that it is quite impossible to review them all here. Olver already identified and cited more than fifty "rediscoveries as generalizations" in his [1986a]. We shall give only one example.

In 1973 the Swedish physicist Dan Anderson published a short article [1973] in which he remained within the case of a single independent variable, "discovered" that Noether's theorem, i.e., Hill's version of the first theorem, could be generalized to Lagrangians of an order higher than 1, and proposed applications to higher-order mechanics. As we observed above, even a well-informed reader like Logan could mistake the contents of this article for a "more modern form" of Noether's theorem.

Other authors cite Noether's original paper but nevertheless fail to apply some of the results of her article; in particular, they treat questions for which Noether's second theorem is relevant in ignorance of it. In her article on conservation laws and their applications in global differential geometry [1983], Karen Uhlenbeck gives a detailed and very interesting description of the applications of Noether's first theorem to such problems as the determination of harmonic maps, the search for Einstein metrics and the solution of the Yang–Mills equations. She cites Noether's original article as well as the formulation by Arnold of the simplest form of Noether's first theorem (Arnold [1974], English translation, 1980, pp. 88–90) upon which she may have based her work. She states "Noether's theorem in Lagrangian mechanics," which is to say, for a simple integral where the integration variable is time applied to a first-order Lagrangian, and she gives several applications including one to the case of multiple integrals where time plays a special role. Then comes "the general formulation of Noether's theorem," which is to say, "the correct invariant formulation." She in fact gives an invariant formulation of the first theorem in modern language in the case of multiple integrals, but still for a first-order Lagrangian and for classical symmetries. She does observe that Noether's results permit treating Lagrangians defined in terms of the curvature of the metric of a Riemannian manifold which depend on the second derivatives of that metric, but she omits describing these results because "[her] simplified formalism only covers first-order Lagrangians." Although she states clearly in her introduction that "problems in general relativity [...] were the original motivation for and application of Noether's theorem," she fails to recognize the existence of the second theorem, and she seems to consider applying the first theorem to the case of the group of all diffeomorphisms of the manifold (p. 113), then to the invariance group of the Yang–Mills Lagrangian, but finally, she concludes her treatment of the equations derived by an application of the first theorem with the reflection, "It is disappointing that these are not particularly enlightening equations." Disappointing perhaps, but surely not astonishing in view of Noether's results on invariance under groups depending on arbitrary functions.

Chapter 6
The Reception of Noether's Second Theorem after 1950

While the historical connection between Noether's first theorem and its consequences for classical and quantum mechanics was not emphasized until late in the twentieth century, the connection between the fundamental result constituted by Noether's second theorem—which has been ignored by most authors of articles and books on the calculus of variations—and general relativity was recognized more explicitly among researchers in this area and that connection was generally acknowledged in the literature since 1950. In the gauge theories that have been developed more recently, its role has also been recognized.

Let us explain the terminology that we will use here. What physicists call "global gauge transformations" or "gauge transformations of the first kind" are transformations which depend on one or several parameters, while "local gauge transformations" or "gauge transformations of the second kind" are transformations that depend on one or several arbitrary functions.[1] When one speaks about "gauge transformations," the expression refers in general to the "local gauge transformations." In addition, one must distinguish between "local conservation laws" that are written in differential form, and "global conservation laws" that are obtained by integration on a domain of the independent variables, which is to say, in the case of general relativity, a domain of space-time.[2]

6.1 The Second Theorem and General Relativity

We shall survey some important texts on general relativity which deal with what Noether called "improper" conservation laws.

---

[1] In this context, the meaning of the words "global" and "local" is different from the meaning of those two words as they are used in differential geometry.

[2] Some of the vocabulary of differential geometry which we shall use in this chapter will be described in Chap. 7, *infra*.

Y. Kosmann-Schwarzbach, *The Noether Theorems*, Sources and Studies in the History of Mathematics and Physical Sciences, DOI 10.1007/978-0-387-87868-3_7,
© Springer Science+Business Media, LLC 2011

Bergmann — In his *Introduction to the Theory of Relativity* [1942], Peter Gabriel Bergmann[3] set out the principles of relativistic mechanics, both special and general, as well as of Weyl's gauge invariant theory [1918a,b] without giving any further bibliography, and, in his subsequent publications, he studied new properties of conservation laws in general relativity. In an article on nonlinear field theories [1949], he derived an expression for the strong conservation laws[4] of a generally covariant nonlinear field theory, with no mention of Noether, and he later recalled their construction in the first part of his joint paper with James L. Anderson [1951]—still without citing Noether[5]—as a preliminary to its second and main part that deals with the associated Hamiltonian formulation of generally covariant field theories in the presence of constraints.

In contrast, in the introduction to the article he published with Robb Thomson [1953], Bergmann wrote, "Noether has shown that, in general, conservation laws obtained in such a way [by invariance under a finite-dimensional subgroup of the group of all coordinate transformations] are simply restatements or even special cases of the general conservation laws resulting from the invariance with respect to curvilinear transformations," and he cited her article [1918c]. He and Thomson then observed that "it seems that the angular momentum law could not be obtained by appealing to an invariance argument in the usual theory," and they proposed circumventing that obstacle by constructing, from the energy-momentum tensor suitably defined, a superpotential for the angular momentum of the total field, that is, including both the matter terms and the gravitational terms.[6]

Five years later the reference to Noether's article disappeared, even though his article on conservation laws in general relativity [1958] begins,

> Throughout mechanics and field theories, it is well known that the fundamental conservation laws are related to the universal invariance properties of physical laws [...] The structure of conservation laws in general relativity and in general-relativistic theories differs from that in nonrelativistic and in Lorentz-covariant theories because of the much wider scope of coordinate transformations in general relativity. It was discovered a long time ago that the so-called conservation laws of energy and linear momentum in general relativity [...] which hold only insofar as the field equations of the theory are satisfied, are related to a set of identities, the "strong" conservation laws [...].

At the beginning of the 1970s, Noether reappeared in Bergmann's thinking when he confided to Kimberling,

---

[3] Bergmann (1915–2002), a former student of Frank in Prague, and a professor at Syracuse University from 1947 to 1982, had been Einstein's assistant and collaborator at the Institute for Advanced Study at Princeton from 1936 to 1940.

[4] See, *supra*, Chap. 2, p. 62. Bergmann's strong laws are trivial laws of the second kind in the sense of Olver because the square of the horizontal differential vanishes, thus the divergence of such a 0-form vanishes identically, just as, in Euclidean 3-dimensional space, whenever a vector field is a curl its divergencce vanishes.

[5] The reviewer of this article for *Mathematical Reviews* wrote that "from the invariance properties of the theory a number of identities (generalized Bianchi identities) are derived," but he did not identify them as "the Noether identities."

[6] When the divergence of a tensor density vanishes identically, that density can be written as a curl, at least locally. The superpotentials are quantities that define such a curl.

Noether's Theorem, so-called, forms one of the corner stones of work in general relativity as well as in certain aspects of elementary particles physics [...] A discussion of the consequences [of the principle of general covariance in general relativity] in terms of Noether's theorem (whether explicitly quoted as such or not) would have to include all of the work on ponderomotive laws, *inter alia*.[7]

Finally, in 1976, when the second edition of his 1942 book appeared, he introduced it with a new preface and added two appendices. In the preface he emphasized the importance of Appendix A, in which he provided a new derivation of the laws of motion of rigid bodies according to the rigorous approach that he and his student Joshua N. Goldberg (see *infra*) had developed. This appendix contains a brief introduction to the definition of an invariance group, and then a section entitled "Noether's theorem," where the connection between symmetries and conservation laws is first stated in the Hamiltonian and then in the Lagrangian formalism. Bergmann distinguishes carefully between "Lie groups" and "function groups," observing that these latter groups are called "gauge groups" by physicists. Then he shows that, using his method of surface integrals, the conservation theorems yield significant results in general relativity.

Goldberg 1953, 1980 — In the article [1953] that he drew from his thesis and which developed several of Bergmann's ideas, Joshua N. Goldberg studied the strong conservation laws both in the Lagrangian and the Hamiltonian formalisms. Much later, in his "Invariant transformations, conservation laws and energy-momentum" [1980], he studied the relations between the global and the local forms of the conservation laws of general relativity. Early in his article he cites Noether and then, in section 2, he formulates "Noether's theorem," for Lagrangians which are of the first order only, and in the case of first-order generalized symmetries. Even though he states a single theorem, he distinguishes carefully within its two cases, the one that supposes constant parameters, and the other that supposes parameters that are arbitrary functions, a distinction which corresponds to Noether's two theorems. He shows that the second case yields strong conservation laws, i.e., vector-valued 0-forms whose divergence vanishes identically, independently of the field equations. These general results are then applied to general relativity where the asymptotic conditions imposed on the fields play an important role. Goldberg studies the expressions for the conserved currents that had been proposed by Einstein [1918] and by Komar [1959], and compares their values, especially their limits at infinity, in the space-like or isotropic directions, and finally decides that Einstein's expressions are better even though they are more difficult to manipulate (p. 488).

Fletcher 1960 — In 1960 John George Fletcher, who had been a student of John A. Wheeler, published a well documented article [1960] whose objective was to summarize what was known about local conservation laws in generally covariant theories, to compare the points of view of physicists who had written on the subject, and to elucidate obscure points and especially those that concerned applications of such conservation laws to physics. He used the term "strong conservation law" for a vector-valued 0-form which is the horizontal differential of a 0-form with values in

---

[7] Quoted in Kimberling [1972], p. 142.

the $(n-2)$-forms on the base, where n is the number of independent variables, i.e., the dimension of the base manifold, when the field equations are satisfied. Because Bergmann held that a conservation law defined by a vector-valued 0-form is strong if the divergence of the 0-form vanishes identically, independently of the fact that the Euler–Lagrange equations are satisfied, a strong law in the sense of Fletcher is also a strong law in Bergmann's sense.[8] These strong laws are "improper laws" in the sense of Noether.

In his section 5, Fletcher presents two methods for producing conservation laws. The first, which we have already described, is that of symmetries. He remarks that "the originator of this method is unknown to the present author. It is discussed, for example, by Pauli [1941]."[9] He then describes Noether's first theorem without attributing it to her, but does remark (p. 74, note 26) that Noether proved in [1918c] that, under certain hypotheses, every local conservation law comes from a symmetry. He calls the second method the "commutator method" and attributes it to Jack Heller [1951]. It consists in effect of applying two infinitesimal symmetries to the Lagrangian to obtain a new local conservation law.

His section 6, "Arbitrary function theorem (Noether's theorem)," deals with Noether's second theorem, first summarized as "In cases in which the parameters p^A are completely arbitrary functions of the coordinates [...], S^ν describes a strong law." Here Noether is acknowledged not only in the subtitle of the section, but also in footnote 29 (p. 75), "This was first proved by Noether [1918c]."

In applications, the fact that a conservation law associated with a symmetry is not unique, because it can be modified by the addition of a total horizontal differential while still remaining a conservation law, has consequences that Fletcher explores. He emphasizes that the conservation laws obtained in generally covariant theories are strong laws, and that they can be written in different forms, which permits him to compare the expressions found by Einstein [1918], Christian Møller [1958], and Komar [1959].

Then, he studies the Hamiltonian point of view and presents the relation between the existence of gauge transformations, that is to say symmetries that depend on arbitrary functions, and the existence of initial constraints among the Hamiltonian variables. He states the fact that each symmetry of order σ will yield σ constraints on the Hamiltonian variables among which some may be trivial. He finally sketches a treatment of gauge symmetries in the formalism of Schwinger or, more exactly, in a semi-classical version of Schwinger's formalism where the commutators of quantum theory are replaced by the Poisson brackets.

Trautman 1962 — In his report on conservation laws in general relativity [1962], Trautman intended "to present the relationship between conservation theorems and invariance properties of physical theories with a particular emphasis on the problem of energy in general relativity." In this article he examined the solutions that had been proposed by Einstein, Hilbert and, later, Bergmann, Komar and Møller, and he

---

[8] See Fletcher [1960], p. 70, note 14.

[9] We remarked above that Pauli in his 1941 article treated only two particular cases, and did not refer to Noether's general results (Chap. 4, p. 93).

succeeded in clarifying this difficult question, though—as we already remarked—it remains a subject of research and discussion. First, with two examples, he illustrated the difference between the "weak" conservation laws obtained by an application of Noether's first theorem on the one hand, and the identities corresponding to an invariance under local gauge transformations depending on arbitrary functions, i.e., gauge transformations of the second kind, on the other. In Dirac's theory of the electron, the invariance of the Lagrangian under the one-parameter group of global gauge transformations leads to the conservation of electrical charge, which is a "weak" law, while, if one modifies the Lagrangian to make it invariant under local gauge transformations, that invariance implies an identity. It is by such a modification of the Lagrangian that one introduces gauge fields. In the case of the electron, the Lagrangian is invariant when the electromagnetic field is multiplied by the phase factor, $e^{i\varepsilon}$, where ε is a constant, the case of a global gauge transformation. One seeks to define a new Lagrangian that will remain invariant under the local gauge transformations, i.e., those in which ε is now an arbitrary function of the space-time coordinates. One is then led to define a gauge field whose components, A_α, are necessarily transformed under a change of coordinates just as the coefficients of a connection for the group $U(1)$ are transformed. The Lagrangian thus obtained by the replacement of the partial derivatives by covariant derivatives with respect to the connection (representing the interaction of the gauge field with the field of the particle), to which one adds a term proportional to $F_{\alpha\beta}F^{\alpha\beta}$, where $F_{\alpha\beta}$ are the components of the curvature of the connection (corresponding to the Lagrangian of the free gauge field), is the invariant Lagrangian that was sought. In general, in quantum field theory, gauge invariance permits determining almost completely the Lagrangian of the system. In general relativity, it is $\nabla_\alpha T^{\alpha\beta}$, the contracted covariant derivative—with respect to the Levi-Civita connection of the metric—of the energy-momentum tensor with components $T^{\alpha\beta}$, which vanishes, and not its ordinary divergence. It is then necessary to add to the energy-momentum tensor a "pseudo-tensor," which is assumed to represent the distribution of gravitational energy and momentum, in order to obtain a conservation law with a physical meaning. This introduction necessitates various choices, but the expressions that are obtained as the various sums only differ from the expression that had been found by Einstein by quantities that are the curls of vector fields.

Trautman continued his report with a section entitled "The Noether Theorems" in which he presented Noether's results, and gave a precise reference to [1918c]. Making only a very few hypotheses to simplify the presentation, he distinguished between strong conservation laws, which correspond to invariance groups depending on arbitrary functions, and weak laws corresponding to invariance under the action of a finite-dimensional Lie group. To conclude the section, Trautman proposed a reformulation of Noether's essential conclusion. "Proper weak conservation laws hold only in such theories for which the symmetry group G_p cannot be extended to a general group $G_{\infty q}$ without introducing auxiliary, nondynamical fields."

He then applied the preceding discussion to the concept of energy in gravitation theories. To obtain conserved quantites in general relativity, because it is the sum of the contracted covariant derivatives of the energy-momentum tensor with contravariant components $T^{\alpha\beta}$, and not its divergence, which vanishes, the corresponding mixed tensor with components T^{α}_{β} must be modified by the addition of quantities t^{α}_{β} corresponding to the energy and the momentum coming fom gravity. The difficulties are due to the fact that these quantities do not have a tensorial character, whence their name, "pseudo-tensors." In particular, the pseudo-tensor that had been considered by Einstein [1916a] can be forced to vanish at a particular point by a suitable choice of coordinates, and that leads to the situation where one cannot speak about the local distribution of gravitational energy, but only of conserved quantities obtained by integration on a space-time domain.[10] The invariance group contains an infinity of one-parameter subgroups, each of which is generated by a vector field. The gravitational action is invariant under these one-parameter groups, whence the existence of an infinity of weak conservation laws but, because of the general covariance, each of these conservation laws can be modified in such a way as to be satisfied independently of the field equations, and in this case the conserved tensor density is a curl. Trautman observed that the existence of superpotentials that define such curls is an important feature of conservation laws in general relativity. He also showed that the identities obtained as a consequence of general covariance are the contracted Bianchi identities.

Cattaneo 1969 — The Italian physicist Carlo Cattaneo lectured on conservation laws in general relativity at the International Conference on Gravitation and General Relativity in London in 1965 and, at the invitation of Lichnerowicz, at the Collège de France in Paris in February 1966. In his article on conservation laws [1966], which is the text of his London lecture, he surveys "the main attempts which have been made to establish conservation equations in general relativity," and he asserts that it was "in accordance with Noether['s] famous theorem which associates the classical conservation laws of the Lorentz-covariant theories with the properties of invariance of physical laws" that Bergmann worked on this question. This assertion would have been more correct if Cattaneo had written "theorems" in the plural, because in Bergmann's work it is Noether's second theorem which is important, but he seems to be unaware of its existence. This fact appears clearly in his subsequent publication, "Invariance and conservation" [1969], which contains an analysis of the limitations of the applicability of "Noether's method" ("il metodo di Noether"), by which Cattaneo obviously means her first theorem. In the first part of this article, he derives the conservation laws associated with an invariant Lagrangian in the case of "conventional field theories" ("teorie di campo convenzionali"), i.e., all field theories except general relativity and the various unitary theories combining gravitation with electromagnetism, and he proves that, for a generally covariant Lagrangian on a space-time manifold possessing sufficiently many Killing vector fields, i.e., vector fields leaving the metric invariant, there exists a "metric energy-momentum

---

[10] This had been observed by Weyl in [1918b], 3rd ed., 1919, §32, "Gravitationsenergie. Die Erhaltungssätze," p. 233, 4th ed., 1921, §33, p. 246; English translation, 1922, p. 271; French translation, 1922, p. 237. There is no section on "Gravitational energy. The conservation laws" in the 1st edition (1918) nor in the 2nd, unchanged edition (1919).

tensor" which is conserved.[11] In the second part of the article, he seeks the limits of applicability of "Noether's method," and sets out to show in what sense the metric energy-momentum tensor furnished by Noether's method becomes illusory ("diviene illusorio") in the case of the Einsteinian gravitation theory, and why nontensorial quantities have to be considered in order to obtain physically significant conservation laws. He derives Einstein's gravitational energy-momentum pseudotensor and the corresponding superpotential, then explains that there exist large families of such pseudo-tensors which have been proposed in the literature, a fact related to the existence of "gauge equivalent Lagrangians," i.e., Lagrangians that differ by a divergence and therefore yield the same field equations. While Cattaneo, in this very clearly written article, gives valuable information about the derivation of Noether's first theorem and its application to field theory, on the one hand, and on energy conservation in general relativity, on the other, he does not recognize that Noether had shown in her second theorem how radical was the change from Minkowski space-time to general, curved space-time, and that she had described the phenomenon of "improper" conservation laws, which Bergmann and Trautman later considered.

Olver 1986 — It was around 1985 that Peter Olver undertook a rigorous mathematical presentation of the proof of Noether's second theorem and its converse. We know of only one earlier attempt, besides Komorowski's articles [1968], at modernizing the proof of the second theorem, that of Francisco Guil Guerrero and Luis Martínez Alonso [1980a]. Olver's proof appeared in an article [1986b] after having been summarized in his book [1986a]. He showed that the existence of differential identities among the components of the Euler–Lagrange derivative of a Lagrangian is due to the under-determined nature of the system of Euler–Lagrange equations and that, in this case, the conservation laws that appear are "trivial." In particular, the Euler–Lagrange system of equations is not then a normal system.[12] Olver's rigorous distinction between the two kinds of trivial conservation laws finally permitted a clear understanding of the situation.

6.2 The Second Theorem and Gauge Theories

As has been explained above, Noether's second theorem concerns the groups of symmetries of a Lagrangian which depend on arbitrary functions. These groups are now called "gauge groups," and they give rise to "gauge theories." In fact, today

---

[11] This result generalizes the conservation of the energy-momentum tensor for a field theory on Minkowski space-time whose Lagrangian is invariant under the inhomogeneous Lorentz group. Cattaneo's derivation follows Trautman's article [1957].

[12] A Lagrangian is said to be *normal* if, in the Euler–Lagrange system of equations, one can express the partial derivative of each unknown function which is of maximal order with respect to one of the variables as a function of all the partial derivatives which are of order strictly less with respect to that variable, and of an order inferior or equal with respect to the other variables. For the definitions of "trivial," "trivial of the second kind" (i.e., "strong"), and "weak" conservation laws, see, *supra*, Chap. 2, p. 62, note 27. These "trivial conservation laws" are Noether's "improper laws."

these gauge symmetries are almost always considered to be the automorphisms of a principal fiber bundle which, when the bundle is trivial, i.e., when it is the product of the base manifold and the structure group, may be identified with the maps from the base manifold to the group, thus with elements of the group depending on arbitrary functions. If the structure group is a ρ-dimensional Lie group, such sections are locally determined by ρ scalar functions.

The evolution of the concept of gauge invariance, since it was first proposed by Weyl in 1918, is complex but well known.[13] Its main sources are Weyl's fundamental article [1929] and Pauli's [1941] twelve years later. In quantum mechanics, the multiplicative factor was replaced by a phase factor; the invariance group of scale transformations, that is to say the group of positive real-valued functions, has thus been replaced by the group of functions with values in the group $U(1)$ of complex numbers of modulus 1, which is also the group $SO(2)$ of rotations in the 2-dimensional real plane. Finally, in 1954, Yang and Mills considered groups that were more general than $SO(2)$ and that were noncommutative. Such groups are also called nonabelian. This was the beginning of the "nonabelian gauge theories," for various nonabelian groups.[14] In all these theories in which the invariance groups depend on arbitrary functions, Noether's second theorem is an essential tool.

A treatment of these questions can be found in Bryce DeWitt's course on field theory [1964], where he describes for an audience of physicists several results in differential geometry and in group theory, but does not use the mathematical apparatus of fiber bundles.[15] In this long article the principal center of interest consists

---

[13] See Yang [1986], Doncel *et al.* [1987], O'Raifeartaigh [1997], Scholz [1999b] [2001], and the book edited by Gerardus t'Hooft [2005].

[14] See, e.g., Carmeli, Leibowitz and Nissani [1990], Henneaux and Teitelboim [1992], O'Raifeartaigh [1997], Deligne and Freed [1999] and, for a wide generalization based on the concept of a superpotential, Julia and Silva [1998]. For the state of the Yang–Mills theories in physics in 2005, see t'Hooft [2005]. For a discussion of Noether's theorems that includes a reference to the history of general covariance, and that discusses the physics of gauge theories, see the publications of Katherine Brading and Harvey R. Brown, in particular [2003], and see the references they cite. More material on gauge theories, and on symmetries in general, from the philosophical and epistemological viewpoints can be found in the various essays of the book edited by Brading and Elena Castellani [2003].

[15] To the best of our knowledge, the first publication that formally identified the components of the Yang–Mills field with those of a connection on a principal fiber bundle was by Hélène Kerbrat-Lunc, in a note in the *Comptes rendus* of the Paris Academy of Sciences presented by Lichnerowicz, "Mathematical introduction to the study of the Yang–Mills field on a curved spacetime" [1964]. About the same time, upon reading the 1960 Russian translation of Lichnerowicz's "Global Theory of Connections and Holonomy Groups" [1957], Ludwig Faddeev also realized that "connections and the Yang–Mills fields [...] were one and the same" (Faddeev [1987], English translation, 1995, p. 11). In fact, in 1980 Yang would write, "That gauge fields are deeply related to the geometrical concept of connections on fiber bundles has been appreciated by physicists only in recent years" ([1980a], p. 44; *Selected Papers*, p. 565), and, in his *Selected Papers*, p. 73, he gives some further information about the evolution of his thinking regarding the relations between gauge theories and the absolute parallelism of Levi-Civita, that is, the theory of connections. On the other hand, in the course of his study of the contribution of Weyl to physics (Yang [1986], p. 17), he explains that, for Weyl, gauge invariance was strongly linked to general relativity, and he adds, "Only in the late 1960s did I recognize the structural similarity mathematically of

of the differential identities attached by Noether's second theorem to the infinite invariance groups that DeWitt simply calls "invariance groups."[16] But he only cites
her for her first theorem on the conservation laws.[17]

The "Noether method" plays an essential part in the theory of supergravity, a
gauge theory where the symmetry, called supersymmetry, exchanges bosonic fields
and fermionic fields.[18] Since the Lagrangian is invariant up to divergence, one modifies it by "Noether coupling," and one iterates the procedure. Because the ultimate
result involves a space-time metric, Peter van Nieuwenhuizen says (p. 202) that it
has been known how to apply the iterated Noether method "since 1915," which is
anachronistic in the strict sense of the reference, but he probably means ever since
curved space-time metrics appeared in the general theory of relativity.

The Lagrangians of gauge theory are those for which the system of Euler–
Lagrange equations is not normal,[19] whence the existence of a constraint manifold.
The equations of motion are under-determined, and the general solution of the equations of motion then involves arbitrary functions. By Noether's second theorem, the
invariance of the action under an infinitesimal gauge transformation, that is to say, a
vector field that depends on arbitrary functions, implies identities among the Euler–
Lagrange equations. The manifold of solutions of the equations of motion remains
invariant under the gauge transformations. By passing from the Lagrangian to the
Hamiltonian formalism, one can apply Dirac's theory of constraints and define the
first-class constraints.[20]

In the BRST (Becchi–Rouet–Stora–Tyutin) theory, the first-class constraints are
considered to be new variables, the antighosts, and in this approach, the variables—
fields and antifields, ghosts and antighosts—satisfy canonical commutation relations with respect to an odd Poisson bracket.[21] In the book by Marc Henneaux and
Claudio Teitelboim [1992], in which the only case considered is that of one inde-

---

non-Abelian gauge fields with general relativity and understand that they were both *connections*
mathematically." I can testify to the fact that as late as 1978, Yang could still jest at the coffee break
of the mathematics and physics departments of the State University of New York at Stonybrook
that he did not know what a principal fiber bundle was!

[16] See DeWitt [1964], chapter 3, p. 594.

[17] See *ibid.*, p. 598, note, referring to p. 211 [*sic* for 241] of Noether's article [1918c] in the
Göttinger Nachrichten, 1918. For the possible origin of this error in page number, see, *supra*,
Chap. 5, p. 120, note 42.

[18] See, e.g., van Nieuwenhuizen [1981].

[19] See, *supra*, note 12.

[20] See, e.g., Anderson and Bergmann [1951]. See numerous references on the canonical formulation of relativity theory in Lusanna [1991]. For an introduction to a modern, geometric approach,
see Gotay, Nester and Hinds [1978].

[21] Physicists usually call the odd Poisson brackets antibrackets, while mathematicians call them
Gerstenhaber brackets (because Murray Gerstenhaber introduced them in deformation theory
[1964]) or Schouten brackets (because the Schouten–Nijenhuis bracket of multivector fields in
differential geometry is a prototypical example). See Cattaneo, Fiorenza and Longoni [2006]
for a short survey of graded Poisson algebras with applications to the BRST and BV (Batalin–
Vilkovisky) methods of quantization.

pendent variable (time), the authors prove Noether's identities[22] and they conclude that the field equations are not independent, but they do not refer to any article prior to 1959, and therefore do not list Noether's article in their references. Starting from the Noether identites associated with the infinitesimal gauge symmetries of a Lagrangian theory, Stasheff [1997] has shown by means of deformation theory that, more generally, there exist links between the Lagrangian theory in the presence of infinitesimal gauge transformations that leave the Lagrangian invariant, the BRST theory, and the Batalin–Vilkovisky method of quantization. The fundamental idea, which we can already identify in Henneaux and Teitelboim [1992] and even in Boyer [1967], is that the Noether identities involve functions $r_i^{(\lambda)}$ on the jet bundle, satisfying $\sum_i r_i^{(\lambda)} \psi_i = 0$, where the ψ_i are the components of the Euler–Lagrange derivative of the Lagrangian, and that the $r_i^{(\lambda)}$ can be used to construct a first-order deformation of the Lagrangian. This is the point of departure for Stasheff's theory of "cohomological physics" that interprets physicists' constructions in terms of homological algebra.[23]

In their article, Tom Fulp, Ron Lada and Jim Stasheff [2003] recall the proof of Noether's second theorem in the formalism of jet bundles, and they extend its validity to the case of symmetries that depend on one or several arbitrary functions, not only of the independent variables but also of the dependent variables and their derivatives. They illustrate their theory by calculating the Noether identities for the gauge symmetries of the variational problem associated with the Poisson sigma-model, a field theory defined in differential-geometric terms. Then they show that the Noether identities correspond to the antighosts of the Batalin–Vilkovisky theory for which they spell out the cohomological interpretation.

Articles dealing with the study of the Noether identites in the framework of homological algebra appear regularly in the literature, extending it by generalizing either the nature of the symmetries or the type of equations under consideration, including nonvariational equations.[24]

---

[22] Henneaux and Teitelboim [1992], section 3.1.3.

[23] See, e.g., Stasheff [2005].

[24] See, e.g., Sardanashvily [2005].

Chapter 7
After 1970—Genuine Generalizations

In the preceding chapter, we described work that uses or generalizes Noether's second theorem. Now we shall look at the generalizations of her first theorem that began to appear in the 1970s. After the pioneering work of Trautman [1967], geometric studies of that theorem began to be undertaken. The first such studies consisted of finding an invariant formulation of the first theorem in the framework of the geometry of differentiable manifolds, which is to say, without using local coordinates. This was accomplished for first-order Lagrangians by Goldschmidt and Sternberg [1973]. As was observed above, the passage to a greater number of independent variables and to higher-order Lagrangians was accomplished in papers that are too numerous to be listed here.

These generalizations only treated the case of classical symmetries, the infinitesimal symmetries in the sense of Lie, but not yet that of generalized symmetries which, as we have seen, were already present in Noether's local theory. In fact, if the relativists had always used symmetries that depend on the derivatives of the field variables, such was not the case for the research that dealt with the first theorem. The redirection of interest toward generalized symmetries is due to the essential role they play in the theory of integrable systems which became the subject of intense research after 1970. In this theory, an evolution equation is seen as defining the flow of a generalized vector field, and one of the essential properties of an integrable system is that it admits an infinite sequence of symmetries, generalized vector fields which depend on derivatives of increasingly higher order of the dependent variable. The bibliography of integrable systems and soliton equations is so immense[1] that we cannot discuss this subject here. We must stress the fact that it is impossible to restrict oneself to the ordinary vector fields in this theory, so the introduction of the generalized symmetries was inevitable.

It was also during the seventies that a theory developed, quite independently of Noether's influence, that led toward the formal calculus of variations *à la* Gel'fand–Dickey–Dorfman and toward the geometry of higher-order jet bundles, or more

---

[1] See for example Manin [1978] and the collection of papers by Novikov *et al.* [1981] and, in particular, George Wilson's introduction. A bibliography of the more recent literature would fill tens of pages.

Y. Kosmann-Schwarzbach, *The Noether Theorems*, Sources and Studies in the History of Mathematics and Physical Sciences, DOI 10.1007/978-0-387-87868-3_8,

precisely, of their projective limit, the infinite-order jet bundles à la Vinogradov–Tsujishita. At the same time the exact sequence of the calculus of variations was discovered and studied by various authors, from Dedecker [1975] and Tulczyjew [1975] [1977] to Toru Tsujishita [1982].

7.1 Jet Bundles and Generalized Symmetries

The ideas that permitted the recasting of Noether's theorems in geometric form and their genuine generalization were first of all that of differentiable manifolds, i.e., manifolds of class C^∞, also called smooth manifolds, sometimes manifolds for short, and then the concept of a jet of order k of a mapping, where k is a non-negative integer, defined as the collection of the values of the components in a local system of coordinates of a vector-valued function and of their partial derivatives[2] up to order k, the concept of manifolds of jets of sections of a fiber bundle, and finally of jets of infinite order. The manifold of jets of infinite order of sections of a fiber bundle is not defined directly but as the inverse limit (also called the projective limit) of the manifolds of jets of order k, as k tends to infinity. A vector field tangent to a fiber bundle is said to be vertical if it is projectable onto the base manifold, and its projection vanishes everywhere, i.e., if it is tangent to the fibers of the bundle. More precise definitions of these terms may be found in the literature. We shall only mention the roles played in these developments first by Ehresmann, for finite-order jets [1951], and then, after 1970, by Vinogradov and his collaborators[3] and, finally, by Tsujishita [1982] among many others.

Once these concepts were formulated, it became possible to define generalized symmetries rigorously. Olver, in the notes to chapter 5 of his book [1986a], brought up to date in a second edition in 1993, sketched a brief history of generalized symmetries. They were indeed first discovered and used by Noether, while earlier, partially successful attempts at generalizing point transformations had been made by Lie and by Albert V. Bäcklund (1845–1922) in the 1870s. Lie defined the contact transformations but did not succeed in generalizing them to transformations involving derivatives of order higher than the first, while Bäcklund, who worked with transformation groups and not with Lie algebras of infinitesimal transformations, showed that indeed there did not exist any transformations of what are now called the jet spaces besides prolongations of point or contact transformations. In fact, the generalized vector fields do not generate one-parameter groups of transformations of either the bundle under consideration or any bundle of its jets of sections of finite order, but they give rise to local one-parameter groups of transformations of function spaces which are obtained by the integration of evolution partial differential equations. It was only with Lie's infinitesimal methods that Noether's generalization was possible.

---

[2] The derivative of order 0 of a function is just the function itself.

[3] Vinogradov [1977] [1979] [1984a,b], Krasil'shchik and Vinogradov [1997].

The rediscovery of generalized vector fields was due to several authors, working independently: Harold H. Johnson (1929–2009) in 1964, who called them "a new type of vector field," Robert Hermann [1965], Heinz Steudel in a series of articles published between 1962 and 1973,[4] Zaur V. Khukhunashvili in little-known articles [1968] [1971], cited by Ibragimov in a recent book, that aim at determining all the symmetries of certain equations of physics, and then Anderson,, Kumei and Wulfman [1972]. Alone among these authors, Steudel, in each of his articles [1965], [1966] and [1967], referred to Noether, but even he does not seem to have recognized the full generality of her theorems.

The prehistory of the nearly contemporaneous articles of Johnson [1964a] and Hermann [1965] is involved but can be reconstructed by means of the summaries that each wrote of the other's publication for *Mathematical Reviews*. Johnson, in his introduction to [1964a], presented this new notion as an example of what Hermann had defined at approximately the same time under the name "tangent vector field on a function space," cited mimeographed notes written by Hermann in 1961, and referred to a paper by Hermann as "in print." Hermann, in the article that was published slightly later [1965], cited Johnson's published paper (p. 302). It was Johnson, a geometer, who, in his articles [1964a,b], introduced a Lie bracket on the vector space of generalized vector fields, and proposed applications to the study of the symmetries of differential equations.

Anderson, Kumei and Wulfman [1972] announced that they had introduced infinitesimal transformations that were more general than those of Lie and Ovsjannikov, and they applied them to the determination of the symmetries of the hydrogen atom. Independently, since around 1967, Ovsjannikov and his student Ibragimov were studying the question of the invariance of differential equations and, in 1976, Ibragimov introduced the concept of "Lie–Bäcklund transformations." A collaboration of Robert L. Anderson, from the physics department at the University of Georgia in Athens, and Ibragimov, from the University of Novosibirsk, was initiated at a conference that took place in Novosibirsk, and was then continued in a series of exchange visits which were rather rare at the time. Their joint work resulted in a book, *Lie–Bäcklund Transformations in Applications* [1979] which became quite influential in the development of the theory of soliton equations. The term, Lie–Bäcklund transformations, was rapidly adopted even though it lent itself to a certain confusion. Strictly speaking, these transformations were not a special case of Bäcklund transformations, nor were Bäcklund transformations a special case of Lie–Bäcklund transformations.[5] More recently, Bluman and Kumei [1989] developed a theory of the symmetries of differential equations that included generalized symmetries.

It was Vinogradov [1977] [1979] who showed that generalized vector fields were nothing other than ordinary vector fields on the bundle of jets of infinite order of sections of a bundle. Then it was both necessary and natural to introduce forms

---

[4] Steudel's articles influenced Śniatycki [1970], and were later cited in an important article by Guil Guerrero and Luis Martínez Alonso [1980b] in which they used generalized variational derivatives to derive the conservation theorems in the case of generalized symmetries.

[5] See Kosmann-Schwarzbach [1979], but when that article was written we did not yet know that the concept of generalized symmetry was already present in Noether.

on this jet bundle. Conservation laws then appeared as a special type of $(n-1)$-forms where n is the dimension of the base manifold, what we call "vector-valued 0-forms" by analogy with the axial vectors in Euclidean 3-dimensional space. The divergence operator may be interpreted as a horizontal differential, which is to say, one that acts on the variables which are the local coordinates of the base, in other words, on the independent variables. One can then introduce an invariant concept of the variational symmetry of a Lagrangian, that is, a generalized symmetry up to divergence.[6]

In fact, Vinogradov[7] considers the inverse limit $J^\infty(F)$ of the bundles of finite-order jets of sections of a bundle $F \to M$. Functions on this infinite-dimensional bundle can be identified with scalar differential operators on F. The vector fields on $J^\infty(F)$ are the derivations of the ring of functions, vector fields whose bracket he calls the Jacobi bracket.[8] The algebra of differential forms is the union of the algebras of differential forms on the bundles of finite-order jets.

There exists an integrable distribution on $J^\infty(F)$, called the Cartan distribution, defined as the tangent space to the graphs of the prolongations of sections of F. Its dimension is the dimension, n, of the base manifold M, and it is locally generated by the total derivations, the vector fields D_i, $i = 1,\ldots,n$, with local coordinate expressions

$$D_i = \frac{\partial}{\partial x^i} + u^\alpha_{I,i} \frac{\partial}{\partial u^\alpha_I}.$$

Every vector field on the base manifold can be uniquely lifted to the Cartan distribution, whence a connection, called the Cartan connection, which is flat. Any tangent vector to $J^\infty(F)$ at a point can be uniquely decomposed as the sum of a vertical vector and a vector belonging to the Cartan distribution.[9] Dually, the cotangent space to $J^\infty(F)$ at a point is the direct sum of the Cartan forms—which vanish on the Cartan distribution—and the horizontal forms. The space of Cartan forms is generated by the $du^\alpha_I - u^\alpha_{I,i} dx^i$, while the space of horizontal forms is generated by the dx^i, whence a bigrading of the space of forms, and the bicomplex of the vertical and the horizontal differentials, such that

$$d_v f = \frac{\partial f}{\partial u^\alpha_I}(du^\alpha_I - u^\alpha_{I,i} dx^i), \qquad d_h f = D_i f\, dx^i.$$

Thus, in Vinogradov's terms, a Lagrangian is a horizontal n-form, and an equivalence class of Lagrangians is a cohomology class for d_h. Applying the vertical

---

[6] By definition, a generalized vector field X is a variational symmetry of a Lagrangian L if the Lie derivative of L by X is equal to the divergence of a vector-valued 0-form. (We remark that Olver in his book [1986] uses the term "variational symmetry" in a more restricted sense.) It follows that X is a variational symmetry of L if and only if its vertical representative is a variational symmetry.

[7] See, e.g., Vinogradov [1984a].

[8] Other authors (Gel'fand and Dickey [1975]) simply call it the Lie bracket, while Kosmann-Schwarzbach [1980] calls it the vertical bracket.

[9] See, *supra*, Chap. 2, p. 58, for the role of the vertical generalized vector fields in Noether's proof of her first theorem.

differential to the given horizontal n-form yields the Euler–Lagrange differential of the Lagrangian.[10]

If $\mathscr{D}$ is a formally integrable differential operator, the space $\mathscr{E}$ of solutions of the infinite prolongation of $\mathscr{D}$ is a "submanifold" of $J^\infty(F)$. A conservation law for $\mathscr{D}$ is then an equivalence class of horizontal $(n-1)$-forms v on $\mathscr{E}$ such that $d_h v = 0$, modulo the d_h-trivial forms, that is to say, those forms that are the image under the horizontal differential of an $(n-2)$-form on $\mathscr{E}$. (We referred, *supra*, to these horizontal $(n-1)$-forms as vector-valued generalized 0-forms.)

These considerations yield a cohomological formulation of the conservation laws which is mathematically satisfying, although considerably distant from the concept of conserved quantity as it was formulated in the classical works on mechanics.

7.2 Characteristics of Conservation Laws and the Converse of the First Theorem

After 1970, various formulations of Noether's first theorem and its converse began to appear, and they broadened its range of applications because they included such concepts as weak variational symmetries.[11] Other converses had been formulated a few years earlier as a result of the controversy over the "zilch tensor."[12] Steudel [1965] [1967] had shown that the new conservation laws for the Lagrangians considered by Lipkin [1964] could be obtained from Noether's first theorem. Then Tulsi Dass [1966] extended this result to other Lagrangians, and formulated a converse of Noether's first theorem for Lagrangians that did not depend explicitly on time, and that was further generalized by Steudel in a short article in the *Annalen der Physik* [1967].

The concept of a characteristic of a conservation law was introduced independently by Gel'fand and Dickey [1975] in the case of one independent variable, then by Magri [1978], Manin [1978], and by Martínez Alonso [1979]. It was Olver [1986a] who systematized the applications of this concept. A vertical generalized vector field is called a characteristic of a conservation law for the Lagrangian L if $EL(X)$ is a divergence, where EL is the Euler-Lagrange differential of L.[13] If one considers only the vertical vector fields, one immediately obtains the following

---

[10] The Euler–Lagrange differential maps vertical vector fields to n-forms on the base manifold, i.e., it is an element of $\mathrm{Hom}_{\mathscr{F}(F)}(\kappa, \wedge^n T^*M)$, where $\mathscr{F}(F)$ is the space of functions on $J^\infty(F)$ and κ is the $\mathscr{F}(F)$-module of the vertical vector fields on $J^\infty(F)$.

[11] See, for example, Candotti, Palmieri and Vitale [1970] and Rosen [1972].

[12] Numerous physicists published research papers that treated the question of the nature of conservation laws for electrodynamics without referring to Noether and, among them, Daniel M. Lipkin [1964] made "the unexpected discovery of six new conservation laws," which constituted "a source of a mathematical *embarras de richesses*." For lack of a physical interpretation of these laws, he called them the components of the "zilch." The subject was discussed by several physicists, and there were rather heated exchanges of opinions regarding the meaning of the "zilch tensor."

[13] See Olver [1986a], chapter 4, or Kosmann-Schwarzbach [1985], section 3.

form of Noether's theorem with its converse: A vertical generalized vector field is a variational symmetry of a Lagrangian if and only if it is the characteristic of a conservation law for that Lagrangian.

One then introduces equivalence relations on the symmetries—a symmetry is trivial if it vanishes on the solutions of the Euler–Lagrange equations— and on the conservation laws—a conservation law is trivial of the first kind if it vanishes on the solutions of the Euler–Lagrange equations, which of course implies that its divergence vanishes on those solutions as well, or trivial of the second kind if the conservation law itself, which is an $(n-1)$-form, is the horizontal differential of an $(n-2)$-form, in which case its divergence vanishes automatically. One then obtains, for those Lagrangians that are normal, a one-to-one correspondence between equivalence classes of variational symmetries and equivalence classes of conservation laws.

As an example of a subtle geometric study, we can cite the work of Kolář who, as early as 1984, produced a really original study of Noether's fundamental identity, which he did not attribute to her, on fiber bundles, $\delta\lambda = DM_S + E$. Here λ is a Lagrangian, $\delta\lambda$ is the Lie derivative of the Lagrangian, E represents the variational derivative of the Lagrangian, D is the divergence operator, and consequently, M_S is a conservation law associated with the symmetry that satisfies $\delta\lambda = 0$. He showed that in fact, in the general case of a multiple integral and a Lagrangian of order 2 or greater, such an M_S is not uniquely determined but is defined globally by the choice of a torsionless linear connection, S, on the base manifold of the fiber bundle.

In their book [1993], Kolář, Peter W. Michor and Jan Slovák state the above cited identity, refer to the original article for a proof, and then deduce a corollary that they call the "Higher order Noether–Bessel-Hagen Theorem" which can be stated as follows. A generalized vector field is a symmetry for the Euler–Lagrange operator defined by a Lagrangian if and only if the variational derivative of the Lie derivative of the Lagrangian with respect to this vector field vanishes.[14] While Noether's first theorem and its generalization by Bessel-Hagen could be easily deduced from Kolář's identity, the authors preferred to give, without any historical justification, the name "Noether–Bessel-Hagen" to a corollary from which the conservation laws so fundamental to her theorem had disappeared, just the contrary of the many cases where her name did not figure in theorems that were in fact special cases of hers.

7.3 The Formal Calculus of Variations

In Moscow, the school of Gel'fand developed the theory of the formal calculus of variations which would become the basis of the study of soliton equations. It consists of a purely algebraic version of the calculus of variations in which the functionals defined by integrals on space or space-time domains are replaced by

---

[14] Kolář, Michor and Slovák [1993], proposition 49.3, p. 388, and corollary, p. 390. The name attributed to this corollary comes from Trautman ([1967], p. 258) who had called an equation "the Noether–Bessel-Hagen equation," an adequate label, adopted by Krupka ([1973], p. 57).

equivalence classes such that two functionals are equivalent if their integrands differ by a total differential. In this theory, the equivalence classes—and not the individual functionals—are called "functionals." The most notable contributions were those of Gel'fand and Dickey [1975] [1976], of Manin [1978], and of Gel'fand and Dorfman[15] from 1979 to 1982.[16] Since some of the research in this domain by the then young Boris Kupershmidt could not be published in the Soviet Union at that time, Manin published a large part of it in his long article [1978]. It was only somewhat later that Kupershmidt's entire work could and did appear under his own name [1980], and it provides a detailed account of both the Lagrangian and the Hamiltonian formalisms in the general setting of fiber bundles, that includes a "Formal Noether Theorem."

These publications were related to what was later called the exact sequence of the calculus of variations, whose origin lies in the work of Vito Volterra (1860–1940) and Helmholtz in the nineteenth century. In 1887, Volterra proved that the second derivative of a functional is symmetric.[17] In modern terms this means that the linearized operator or linearization—also known as the Fréchet derivative,[18] or the Gâteaux or Hadamard derivative—of an Euler–Lagrange operator is self-adjoint. In the same year, Helmholtz [1887] showed that equations defined by a self-adjoint linear operator are the Euler–Lagrange equations of a Lagrangian. Helmholtz's theorem is a first answer to the "inverse problem of the calculus of variations" which consists in characterizing those equations (or systems of equations) that are the Euler–Lagrange equations of a Lagrangian. Such equations are said to derive from a Lagrangian or to be variational equations. Helmholtz's result was only a first step toward the solution of the inverse problem because it was only valid of course for functionals defined on Euclidean space, not on arbitrary differentiable manifolds, and because he only considered linear operators. Extensions of Helmholtz's theorem have since been the subject of very many papers. The inverse problem of the calculus of variations for nonlinear operators has been solved by replacing the condition of self-adjointness for the operator itself by that for its linearization, i.e., its Fréchet derivative. Toward the end of the 1960s, Tonti [1969] had identified sufficient conditions for a nonlinear differential operator to derive from a Lagrangian. In the late 1970s, the inverse problem of the calculus of variations became part of a much larger theory, the construction and study of the exact sequence of the calculus of variations included in the variational bicomplex. In fact, the inverse problem is related to just the first term in the infinite sequence of terms in this exact sequence. This construction was the work of numerous mathematicians, but mainly

---

[15] Irene Dorfman, who died in Moscow in 1994, made remarkable contributions to the algebraic theory of integrable systems.

[16] See Dorfman [1993].

[17] Volterra [1887], pp. 103–104; *Opere matematiche*, vol. 1, p. 302. Volterra proved this result again in his book [1913], p. 47. His term for "functionals" is the very explicit "functions that depend on other functions" ("funzioni che dipendono da altre funzioni"), which became in French "fonctions de ligne."

[18] Maurice Fréchet (1878–1973) was a French mathematician whose work dealt with the formulation of abstract topology, analysis and probability.

Dedecker[19] [1975], Tulczyjew [1975] [1977], Vinogradov [1977] (who called it the $\mathscr{C}$-spectral sequence), Floris Takens[20] [1979], Ian M. Anderson and Thomas E. Duchamp [1980], and finally, Tsujishita [1982].[21] This theory is in fact an extremely vast generalization of Noether's first theorem. For example, Takens [1979] showed that, under appropriate conditions, a system of differential equations for which there exists an associated conservation law for each infinitesimal symmetry is necessarily a system of variational equations. This result was itself generalized in a series of articles by Ian Anderson and Juha Pohjanpelto beginning with [1994].

We shall now show that, as research on the exact sequence of the calculus of variations progressed, genuine generalizations of Noether's first theorem were discovered.

7.4 Symmetries and Conservation Laws for Nonvariational Equations

Among the genuine generalizations of Noether's correspondence between symmetries and conservation laws, the one that concerns nonvariational equations appears in the work of several authors that we have already cited. The essential idea can already be found in De Donder's book [1935], as we have mentioned above (p. 100), and would be further developed by Magri [1978] and Vinogradov [1984a]. To see why their reults are a generalization of Noether's first theorem, we must first recall the classical result of Volterra cited above, that the linearization of an Euler–Lagrange operator is self-adjoint. The relation that Noether established between the symmetries of a Lagrangian and the conservation laws for the associated Euler–Lagrange operator is replaced in this generalization by a relation between characteristics of conservation laws for a (not necessarily variational) operator and the generalized vector fields satisfying a condition for the adjoint of its linearization. In the case of an Euler–Lagrange operator, one recovers Noether's first theorem because its linearization is self-adjoint. The search for conservation laws of a given equation is important in practice, and, if the equation is not variational, this generalized Noether correspondence permits replacing that search by the more practical search for generalized vector fields satisfying a condition similar to—but different from—the condition to be satisfied by the generalized symmetries of the equation.

Hermann — Robert Hermann refers to Noether neither in his book *Differential Geometry and the Calculus of Variations* [1968], where he devotes a long chapter to the

---

[19] Paul Dedecker (1921–2007) wrote his doctorate on the inverse problem of the calculus of variations at the Université Libre de Bruxelles in 1948 and had a distinguished research career in the fields of algebraic topology and category theory. He was a professor at the University of Lille from 1963 to 1971.

[20] The Dutch mathematician Floris Takens (1940–2010) was a professor at the University of Groningen whose research was mainly on dynamical systems.

[21] See Ian Anderson's "Introduction to the variational bicomplex" [1992].

symmetries of variational problems, nor in the subsequent book [1970a], where he proves the formula for the conserved charge density associated with a one-parameter group of symmetries in field theory (p. 82). There he also supplies a criterion[22] for the association of a conserved current to a second-order linear differential operator that is invariant under a one-parameter group of symmetries, and he observes that this criterion is satisfied if the differential operator in question derives from a Lagrangian, necessarily a first-order Lagrangian that is quadratic in the fields because the operator is assumed to be linear. In fact, one also knows, by the theory of the inverse problem of the calculus of variations, that this condition is necessary, at least locally. This is thus a real but very limited generalization of Noether's theorem to operators whose linearization is not self-adjoint. In the same year as the publication of this result, Hermann used the expression "Noether's theorem" in another book [1970b], p. 157, so he may have become aware of her article, probably indirectly, in the course of the work that we have just described. Finally, an explicit reference to Noether's "classic article" appears on pages 194 and 199 of his book [1973], though the reference is not included in the list of references.

Magri — In an article in Italian [1978], Franco Magri set out clearly the relation between symmetries and conservation laws for nonvariational equations, but still without treating the most general situation of operators defined on manifolds. He showed that, because finding the conservation laws for a differential operator amounts to finding their charateristics, which he calls "integrating factors" ("operatori integranti"), in order to determine the conservation laws it is sufficient to characterize the kernel of the adjoint of the linearization. This is the fundamental result that permits extending the algorithmic search for conservation laws to equations that do not necessarily arise from a variational problem. Magri explicitly presents the contents of his paper as an extension of "Noether's theorem," but he does not include a reference to Noether's article. Because Magri's article was neither translated nor published in an international journal, its rich results remained largely unknown outside a very small circle.

We shall now sketch Magri's operatorial formulation of Lagrangian theory and of its generalization. When considering an equation $\mathscr{D}(u) = 0$, for unknown functions, $u = (u^\alpha(x^i)), i = 1, \ldots, n, \alpha = 1, \ldots, p$, he defines the Fréchet derivative or linearized operator of $\mathscr{D}$,

$$(V\mathscr{D})_u(v) = \frac{d}{dt}_{|t=0} \mathscr{D}(u+tv),$$

an operator that is linear in v. Explicitly,

$$(V\mathscr{D})_u(v) = \frac{\partial \mathscr{D}}{\partial u^\alpha} v^\alpha + \frac{\partial \mathscr{D}}{\partial u^\alpha_i} v^\alpha_i + \cdots + \frac{\partial \mathscr{D}}{\partial u^\alpha_I} v^\alpha_I.$$

If X is a vector field on the functional space of the u's, one can consider $(V\mathscr{D})_u(Xu)$, denoted by $(V\mathscr{D}(X))(u)$. An infinitesimal symmetry of $\mathscr{D}$ is a vector field X on the functional space which leaves the space of solutions of $\mathscr{D}(u) = 0$ invariant, that is

---

[22] Hermann [1970a], chapter 5.3, p. 167.

to say,

$$V\mathscr{D}(X) = 0$$

whenever $\mathscr{D}(u) = 0$. Magri defines an operator Y to be a characteristic of the symmetry X of $\mathscr{D}$ if

$$V\mathscr{D}(X) = \langle Y, \mathscr{D} \rangle,$$

where $\langle Y, \mathscr{D} \rangle = \mathscr{D}_\alpha Y^\alpha$ if $\mathscr{D} = (\mathscr{D}_\alpha)$ and $Y = (Y^\alpha)$.

By definition, when β is a vector-valued operator, $\beta = (\beta^i)$, the divergence of β is the scalar operator, $\operatorname{div} \beta = \sum_i D_i \beta^i$, where the D_i's, $i = 1, \ldots, n$, are the total derivations, $D_i = \dfrac{\partial}{\partial x^i} + u_{I,i}^\alpha \dfrac{\partial}{\partial u_I^\alpha}$, and a conservation law for $\mathscr{D}$ is a vector-valued operator, β, such that

$$\operatorname{div} \beta = 0$$

whenever $\mathscr{D}(u) = 0$.[23] An integrating factor—elsewhere called a characteristic or a generating function—of a conservation law β for $\mathscr{D}$ is an operator Y such that

$$\langle Y, \mathscr{D} \rangle = \operatorname{div} \beta.$$

In fact, if Y satisfies this relation, $\operatorname{div} \beta$ vanishes on the solutions of $\mathscr{D}(u) = 0$, and β is a conservation law for $\mathscr{D}$.

When L is a functional of order k, $L = L(x^i, u^\alpha, u_i^\alpha, \ldots, u_I^\alpha)$, where I is an unordered multi-index of length $|I| = k$, $(i_1, \ldots, i_k)$, $1 \le i_1 \le n, \ldots, 1 \le i_k \le n$, the Euler–Lagrange differential, EL, of L is $EL = \sum_J (-1)^{|J|} D_J L$. To determine the conservation laws for an operator $\mathscr{D}$, condition $\langle Y, \mathscr{D} \rangle = \operatorname{div} \beta$ can be replaced by condition $E \langle Y, \mathscr{D} \rangle = 0$, since $EL = 0$ if and only if L is a divergence. It is easy to show that

$$E \langle Y, \mathscr{D} \rangle = (VY)^*(\mathscr{D}) + (V\mathscr{D})^*(Y).$$

(If Δ is a linear operator, Δ^* denotes its formal adjoint.) Therefore a necessary condition for Y to be the characteristic of a conservation law is

$$(V\mathscr{D})^*(Y) = 0$$

whenever $\mathscr{D}(u) = 0$. By Volterra's result, if $\mathscr{D} = EL$ for a Lagrangian L, then $(V\mathscr{D})^* = V\mathscr{D}$, and therefore the preceding considerations prove that every characteristic of a conservation law for EL is a symmetry of EL. But these considerations yield more than this formulation of Noether's first theorem. They actually show that searching for the restriction of the kernel of $(V\mathscr{D})^*$ to the solutions of $\mathscr{D}(u) = 0$ is an algorithmic method for the determination of the conservation laws for a (possibly nonvariational) equation $\mathscr{D}(u) = 0$.

Fokas and Fuchssteiner — Athanassios Fokas and Benno Fuchssteiner, in their important article [1980] on integrable systems, introduce the concept of a "conserved covariant," define a "Noether operator" as one that "maps conserved covariants

---

[23] More precisely, if β is an operator with values in the $(n-1)$-forms on the n-dimensional manifold of the independent variables, $\operatorname{div} \beta$ is an operator with values in the n-forms.

onto symmetries," and provide a form of generalized Noether correspondence in an infinite-dimensional Hamiltonian setting related to the then recent work on bi-hamiltonian equations by Gel'fand and Dorfman, Magri, and Olver.

Vinogradov — During his student years, Vinogradov had occasion to consult the books by Polak [1959] and Gel'fand and Fomin [1961] in both of which he encountered the Noether theorems. In the early 1970s he tried to understand the profound nature of the Euler–Lagrange derivative, and since then he began to introduce the edifice of the bundle of infinite-order jets of sections of a fiber bundle over a manifold, which construction permitted him to define generalized vector fields and their bracket, the integrable distribution that he called the Cartan distribution, and the concept of a generating function of a conservation law,[24] while making explicit the role of the condition of being self-adjoint for nonlinear differential operators. In a later article [1984a], he generalized Noether's first theorem, showing that the conservation laws of an operator are associated by a very simple relation to the generalized vector fields determined by a condition on the adjoint of the linearization of that operator, a condition which is the analogue of the invariance relation. Together with his collaborators and students, Vinogradov has proposed many applications of this theorem which show that it is a powerful generalization of Noether's results.

Tsujishita — In a very rich article [1982], Tsujishita unified the study of the bicomplexes which enter into the calculus of variations and into the theory of foliations, and he interpreted the associated spectral sequence geometrically. Noether's first theorem for a normal system of Euler–Lagrange equations[25] belongs to a very important special case of his theory.

Olver — Olver's *Applications of Lie Groups to Differential Equations* [1986a], which we have mentioned several times, presents the most accessible exposition of Noether's two theorems in generalized form. He presented enough background information and formulated the results in such a way that they are immediately applicable to the search for the symmetries and conservation laws of differential equations. This book contains the complete theory of both Noether's theorems, together with many examples of their applications, and a chapter on the complex associated with the calculus of variations. Olver includes numerous important results that were scattered throughout a vast literature, while proving several new results, such as his general theorem 4.17 on the reduction of order of Euler–Lagrange equations. He also emphasized the connection between non-normal systems of equations and Noether's second theorem. He states and proves the general form of Noether's first theorem, and presents a detailed study of a modern form of her second theorem. His book contains a thorough treatment of Lagrangian and Hamiltonian systems, as well as that of the variational complex and bihamiltonian systems.

Is Olver's presentation geometrical? In fact, he first presents all the necessary tools of differential geometry and Lie group theory. Then, in order to make all the

---

[24] Vinogradov [1977] [1979]. Vinogradov's "generating functions" coincide with Olver's "characteristics" in [1986a], and therefore also with Magri's "integrating factors" in [1978].

[25] For normal systems, see, *supra*, Chap. 6, p. 129, note 12. Tsujishita calls such systems of partial differential equations "pseudo-Cauchy–Kovalevski systems." See Olver [1986b].

important results of Lagrangian theory, in particular the determination of conservation laws, directly applicable, he chooses to consider only trivial vector bundles over vector spaces. He may therefore use when necessary global coordinates, both for the independent and the dependent variables. Therefore Olver's book does not contain a complete geometrization of Lagrangian theory—that would involve manifolds instead of vector spaces—but his treatment is so clear and so thorough that a passage to a completely geometric formulation is straightforward from a reading of his text.[26]

7.5 At the end of the twentieth century

The articles and books that appeared after Olver's [1986a] in which one or the other of Noether's theorems of 1918 is used are extremely numerous and varied. Many recall the contents of part of her results in modern language. We single out only two important publications because they constitute a lucid tribute to her contribution to the modern theory of mechanics and field theory.

Gregg Zuckerman [1987] proposed a new universal conservation law in an invariant formalism, and, after pointing out the role of Hamilton's least action principle, he emphasized that "Noether's principle—a continuous symmetry of the action leads to a conservation law—is equally basic." He continued, "E. Noether's famous 1918 paper, 'Invariant variational problems' crystallized essential mathematical relationships among symmetries, conservation laws, and identities for the variational or 'action' principles of physics. [...] Thus, Noether's abstract analysis in [1918c] continues to be relevant to contemporary physics (as well as to applied mathematics (see Olver [1986a]))."

In the comprehensive study that has been cited above, Deligne and Freed [1999] gave a systematic exposition of the Noether theorems[27] in a very general formalism which uses Zuckerman's approach. In addition to those theorems, in the glossary of this book one finds the terms "Noether current" and "Noether charge." By the end of the twentieth century the importance of these concepts and their attribution to Noether finally became a commonplace.

---

[26] In Kosmann-Schwarzbach [1985], we presented a geometrical approach to the Lagrangian and Hamiltonian formalisms, based on the consideration of the finite-dimensional jet bundles of a vector bundle, the generalized vector fields defined in terms of sectional differential operators and their vertical representatives. We introduced the bigraded space of generalized differential forms and we formulated Noether's first theorem in a general geometric setting, using a deep result of Kupershmidt [1980].

[27] Deligne and Freed [1999], chapter 2, pp. 153–190, and in particular sections 2.6 to 2.9.

Conclusion

Several facts now seem clear in light of the documents which we have studied. The *Invariante Variationsprobleme* was not followed in Noether's work by any research with the same orientation, and her subsequent, important contributions in the area of pure algebra—the general theory of ideals and of the representations of algebras—apparently overshadowed it even in her own opinion, and certainly in that of her contemporaries, Einstein, Weyl, and Wigner. In fact, the *Invariante Variationsprobleme* answered questions that had been suggested by Hilbert and Klein, and it was therefore, in some sense, created for the occasion, circumstantial, prompted by the Göttingen debate on the general theory of relativity. When Alexandrov wrote a eulogy of Noether in 1935,[1] on the one hand he affirmed that, together with her earlier work on algebraic invariants, her work on differential invariants which included the two theorems of the *Invariante Variationsprobleme* would have been sufficient to establish her reputation as a first-class mathematician, and that they constituted a contribution to mathematics in no way inferior to the famous work of Sofia Kovalevskaya, and on the other hand, he argued that

> Emmy Noether herself is partially responsible for the fact that her work of the early period is rarely given the attention that it would naturally deserve [...] She herself was ready to forget what she had done in the early years of her scientific life, since she considered those results to have been a diversion from the main path of her research, which was the creation of a general, abstract algebra.

Indeed the value of her contributions to mathematical physics was fully recognized in Göttingen by Klein, especially at the time of the reprinting of his 1918 articles in his *Gesammelte mathematische Abhandlungen*, by Hilbert, though somewhat less generously, and to a certain extent by Weyl [1918b][1935a], and by Einstein, in his correspondence, though he did not cite her results in his scientific articles. But the fundamental importance of her contributions did not appear until much later.

When searching for factual reasons for the nearly complete lack of appreciation of her work of 1918, the facts that Noether was a woman and that she was Jewish come to mind. Is is well known that while the universities and professorships

---

[1] Alexandrov [1936]. See, *supra*, Chap. 3, p. 79.

Y. Kosmann-Schwarzbach, *The Noether Theorems*, Sources and Studies in the History of Mathematics and Physical Sciences, DOI 10.1007/978-0-387-87868-3,
© Springer Science+Business Media, LLC 2011

were in theory open to Jews, the careers of Jewish scientists in Germany faced more obstacles than did those of equally gifted gentile colleagues, and the obstacles to the higher education and academic employment of women are equally well documented.[2] However, Noether's sex and religious confession, and her consequent failure to obtain a prestigious chair in mathematics did not prevent her work in algebra from being immediately recognized and admired. In the case of the 1918 theorems, it may be that Noether's contemporaries, quite naturally, thought that she had worked under the influence of Klein and that it sufficed to read his papers to learn the major lines of her work.

Another fact is that for Klein, "Noether's theorems" were, as his letter to Pauli of March 1921 shows,[3] a very well known collection of results. Some physicists may have thought that for that reason it was not necessary to attribute them to her explicitly. But this is not a satisfactory explanation for the lack of references to her article because, even as a convenient collection of results, the article should have been fully exploited which, as we have shown, was not the case.

One may also suppose that, among the physicists, there were many whose interests were limited to areas where her theorems did not yet seem to apply, and others who lacked the mathematical culture necessary to understand the import of her two theorems. One can even imagine that some of the physicists of the time were leery of the importance that certain pure mathematical theories had acquired. As an example that goes beyond caution to outright hostility, some physicists derisively referred to Lie theory, which was at the heart of the *Invariante Variationsprobleme*, as the *Gruppenpest* ("plague of groups"). This attitude did not completely disappear within the physics community until well after the Second World War. More generally, we can learn of the physicists' opinion of mathematics in an article by the physicist Yang written in honor of the mathematician Shiing-Shen Chern:

> The development of physics in the twentieth century is characterized by the repeated borrowing from mathematics at the fundamental conceptual level [...] Yet it should be emphasized that in each of these cases, the conceptual origin of the physical development was rooted in physics, and not in mathematics. There was in fact, often a certain amount of resistance among physicists to the mathematization of physics.[4]

Yang then cites an amusing letter from Faraday to Maxwell as "a good example of the resistance to the mathematization of physics," describes his own hesitant approach to the geometry of fiber bundles,[5] and concludes that mathematics and physics "have their separate aims and tastes. They have distinctly different value judgments, and they have different traditions. At the fundamental conceptual level they amazingly share some concepts, but even there, the life force of each discipline runs along its own veins." It is therefore not so surprising that it took Noether's mathematics several decades before they were fully accepted by the community of physicists.

---

[2] See, e.g., Tobies [2001].

[3] See, *supra*, Chap. 4, p. 93, and, *infra*, Appendix III, pp. 159–160.

[4] Yang [1980b], p. 250.

[5] See, *supra*, Chap. 6, p. 130, note 15.

It is true that Noether's theorems and the arguments contained in her article provided a key to understanding the nature of conservation laws in general relativity, in particular a basic concept, the conservation of energy. When the physicist Valentine Bargmann (1908–1989) wrote an article on Pauli's contribution to the theory of relativity, he asserted that, when Pauli wrote "his early masterwork," his encyclopedia article of 1921, "the significance of the general covariance of the field equations (notably for the conservation laws) had been clarified by Einstein himself, by the Göttingen school (F. Klein, D. Hilbert, and E. Noether), and by H. A. Lorentz."[6]

In fact it was Noether who was responsible for an important clarification of the problem of energy conservation in general relativity; however this concept remains incompletely understood to this day. In addition, Noether's results did not provide physicists with explanations for other phenomena associated with gravitation. Bessel-Hagen's application of Noether's first theorem to the conformal invariance of Maxwell's equations did in fact answer a question that Klein had asked, but its publication did not arouse the interest of the physicists any more than Noether's article had, and that supports our argument that the lack of reception of Noether's theorems had more to do with the nature of the interests of the mathematical physicists of the time than with the quality of her results or her person.

When her article came into fashion, it did so very gradually. While, already in the 1950s, mathematical physicists working on the theory of general relativity used Noether's second theorem and often knew about the first theorem as well, physicists studying quantum mechanics and quantum field theory mentioned her name only occasionally, in connection with the study of charges, and even then rarely referred to her article. In classical mechanics and in pure mathematics, it was only slowly and a decade later that the number of citations of Noether's article began to increase.[7]

Since wherever a variational principle exists, the determination of the symmetries of the Lagrangian—more generally, of its generalized symmetries, which came to the fore with the development of the theory of integrable systems after 1970—and thus of the corresponding Euler–Lagrange equations, permits obtaining conservation laws which serve in the search for the solutions of these variational equations. That is why the area to which Noether's first theorem is applicable is so wide. In its early years, the study of elasticity, to take only one example, was undertaken independently of the theory of symmetries and so, necessarily, without reference to the Noether theorems. While conservation laws were discovered in the 1950s and 1960s and applied to the study of dislocations and of the diffusion of waves in elastic media, in particular by J. D. Eshelby from 1956 on, their relation with symmetries

---

[6] Bargmann [1960], p. 187.

[7] A rapid survey of the articles tallied in *Mathematical Reviews* which refer explicitly or implicitly to Noether's 1918 article and whose title contains the word "Noether" shows no references before 1950, two in French and one in Slovak between 1951 and 1960, 14 between 1961 and 1970, and then the number per decade increases very rapidly, as does the overall number of mathematical publications. The other articles whose titles contained the word "Noether" dealt either with theorems on different subjects or, for a small number of titles, theorems due to Max Noether or historical studies on Emmy Noether. As for the word *Noetherian*, it is indeed extremely frequent, but this adjective generally refers to rings or algebras or, more recently, to categories or schemes, extrapolations from Noether's work in algebra which are not the subject of this study.

was clarified only after 1980 by the use of Noether's first theorem, when it became possible to systematize the search for conservation laws and to determine new ones associated with generalized symmetries.[8]

Noether's first theorem has been used in numerous theoretical questions concerning partial differential equations, problems of global existence of solutions and problems of stability. Thus on 22 August 1998, in Berlin, at the International Congress of Mathematicians that is held every four years, Cathleen Synge Morawetz, a well-known specialist in the theory of partial differential equations, delivered a lecture entitled "Variations on the Conservation Laws for the Wave Equation."[9] It was long ago recognized that Noether's first theorem has applications in fluid mechanics and, as Blaker and Tavel have shown, in geometric optics [1974]. It is now further recognized that that theorem has important applications to the mechanics of nonholonomic systems.[10] One can cite the generalized formulations of Noether's first theorem and their applications to problems of locomotion in the work of Anthony M. Bloch, P. S. Krishnaprassad, Marsden and Richard M. Murray [1996]. Discrete versions of Noether's first theorem appeared, starting with Logan's "First integrals in the discrete variational calculus" [1973], and continue to appear,[11] and have many applications in the vast field of numerical analysis.

In the second half of the twentieth century, through the contributions of geometers and algebraists, many new domains of pure mathematics developed from Noether's fundamental article. They include the geometric theory of the calculus of variations on manifolds, the definition and determination of the properties of the exact sequence of the calculus of variations, the development of the associated cohomological theory, and the determination of conservation laws for nonvariational equations. Simultaneously, the influence of the *Invariante Variationsprobleme* manifested itself in a large number of areas in mechanics and physics. Thus its history is connected not only with the development of classical mechanics, of general relativity and of the calculus of variations, but also with that of differential geometry, quantum mechanics and quantum field theory, and nonabelian gauge theories. The diversity of fields to which Noether's theorems are now known to be applicable has multiplied the lines of transmission of the information about her original work to the point where few practicing scientists are aware of the full content of her 1918 article. There is still much work to be done to describe the evolution of ideas, of research and even of fashion, according to periods and milieus in these diverse fields, and to establish how Noether's ideas were transmitted in each of them.

---

[8] See Eshelby [1975]. For the first derivation of conservation laws in elasticity theory that was based on symmetries, see Olver [1984], and see Olver [1986a], p. 288, 2nd. ed., 1993, p. 282, for a short historical account.

[9] This lecture was one in the series of the "Emmy Noether Lectures," delivered by a distinguished woman mathematician at each of these congresses.

[10] See, e.g., Krupkova [2009].

[11] See, e.g., Mansfield and Hydon [2001] and Mansfield [2005].

Appendix I
Postcard from Emmy Noether to Felix Klein, 15 February 1918

Verso of the postcard from Emmy Noether to Felix Klein, 15 February 1918
(Niedersächsische Staats- und Universitätsbibliothek Göttingen)

Transcription of the verso of the postcard from
Emmy Noether to Felix Klein, 15 February 1918[1]

[...] also Christoffel & Ricci.

Den Energiesatz habe ich mir erst im einfachsten Fall überlegt, da eine direkte Verallgemeinerung von $f(z_1 \ldots z_\rho, dz_1 \ldots dz_\rho)$ kein einfaches Integral darstellt. Es findet sich nun das n-fache Integral $\delta \int \cdots \int_n f(z_1 \ldots z_\rho, \frac{\partial z_i}{\partial x_\kappa}) dx_1 \ldots dx_n$. Hier wird die „Lagrange'sche Zentralgleichung"

$$\delta f - \frac{\partial}{\partial x_1} \sum_i \frac{\partial f}{\partial \frac{\partial z_i}{\partial x_1}} \delta z_i - \cdots \frac{\partial}{\partial x_n} \sum_i \frac{\partial f}{\partial \frac{\partial z_i}{\partial x_n}} \delta z_i = -\sum \psi_i(z) \delta z_i;$$

ersetzt man nun δz_i der Reihe nach durch $\frac{\partial z_i}{\partial x_1}, \cdots, \frac{\partial z_i}{\partial x_n}$, so geht, wenn f von den x frei ist, δf über in $\frac{\partial}{\partial x_\kappa} f$. Man hat also die n Identitäten

$$\frac{\partial}{\partial x_1} \left(\sum_i \frac{\partial f}{\partial \frac{\partial z_i}{\partial x_1}} \frac{\partial z_i}{\partial x_\kappa} \right) + \cdots + \frac{\partial}{\partial x_\kappa} \left(\sum_i \frac{\partial f}{\partial \frac{\partial z_i}{\partial x_\kappa}} \frac{\partial z_i}{\partial x_\kappa} - f \right) + \cdots \frac{\partial}{\partial x_n} \left(\sum \frac{\partial f}{\partial \frac{\partial z_i}{\partial x_n}} \frac{\partial z_i}{\partial x_\kappa} \right)$$

$$(1) \qquad = \sum_i \psi_i(z) . \frac{\partial z_i}{\partial x_\kappa}; \quad (\kappa = 1, 2 \ldots n);$$

wodurch lineare Kombinationen der ψ als Tensor Divergenzen hergestellt sind.

Das werden für $\psi_i = 0$, n Energiegleichungen, hat man aber *Invarianz*, d.h. nimmt f bei Transformation der x (wobei z in sich übergeht, $\frac{\partial z_i}{\partial x_\kappa}$ sich linear transformiert) f die Transformationsdeterminante als Faktor an, so ist das das Analogon der Homogenität erster Ordnung; dabei verschwinden gerade in (1) die Argumente von $\frac{\partial}{\partial x_1} \cdots \frac{\partial}{\partial x_n}$ identisch, und man hat die n *Identitäten zwischen* den ψ: $\sum \psi_i(z) \frac{\partial z_i}{\partial x_\kappa} = 0$ $(\kappa = 1, 2 \ldots n)$; so daß die ρ Gleichungen $\psi_i = 0$ mit $\rho - n$ äquivalent sind. – Den allgemeinen Fall, wo statt der Skalare z_α Tensoren $g_{\mu\nu}$ stehen, hoffe ich ähnlich erledigen zu können.

Ihre sehr ergebene

Emmy Noether.

---

[1] Manuscript in the Niedersächsische Staats- und Universitätsbibliothek, Göttingen, reproduced on the preceding page. Publication kindly authorized by the Bibliothek.

Translation of the verso of the postcard from
Emmy Noether to Felix Klein, 15 February 1918[2]

[...] as well as Christoffel and Ricci, .

I have considered the energy law first in the simplest case, which represents a direct generalization to $f(z_1 \ldots z_\rho, dz_1 \ldots dz_\rho)$ of a simple integral. One then finds the n-fold integral $\delta \int \cdots \int_n f(z_1 \ldots z_\rho, \frac{\partial z_i}{\partial x_\kappa}) dx_1 \ldots dx_n$. The "central Lagrangian equation" becomes

$$\delta f - \frac{\partial}{\partial x_1} \sum_i \frac{\partial f}{\partial \frac{\partial z_i}{\partial x_1}} \delta z_i - \cdots \frac{\partial}{\partial x_n} \sum_i \frac{\partial f}{\partial \frac{\partial z_i}{\partial x_n}} \delta z_i = - \sum \psi_i(z) \delta z_i;$$

if one now replaces δz_i successively by $\frac{\partial z_i}{\partial x_1}, \cdots, \frac{\partial z_i}{\partial x_n}$, then, if f is independent[3] of the x, δf is transformed into $\frac{\partial}{\partial x_\kappa} f$. One thus obtains the n identities

$$\frac{\partial}{\partial x_1}\left(\sum_i \frac{\partial f}{\partial \frac{\partial z_i}{\partial x_1}} \frac{\partial z_i}{\partial x_\kappa}\right) + \cdots + \frac{\partial}{\partial x_\kappa}\left(\sum_i \frac{\partial f}{\partial \frac{\partial z_i}{\partial x_\kappa}} \frac{\partial z_i}{\partial x_\kappa} - f\right) + \cdots \frac{\partial}{\partial x_n}\left(\sum_i \frac{\partial f}{\partial \frac{\partial z_i}{\partial x_n}} \frac{\partial z_i}{\partial x_\kappa}\right)$$

(1) $$= \sum_i \psi_i(z) \cdot \frac{\partial z_i}{\partial x_\kappa}; \quad (\kappa = 1, 2 \ldots n);$$

by which linear combinations of the ψ are represented as tensorial divergences.

There are thus, for $\psi_i = 0$, n energy equations; however if there is *invariance*, that is to say if under transformation of the x (where z is transformed into itself, $\frac{\partial z_i}{\partial x_\kappa}$ is transformed linearly) f acquires the determinant of the transformation as a factor, it is the analogue of first-order homogeneity; then the arguments of $\frac{\partial}{\partial x_1} \cdots \frac{\partial}{\partial x_n}$ vanish directly *identically* in (1), and one obtains the n *identities among the* ψ:

$$\sum \psi_i(z) \frac{\partial z_i}{\partial x_\kappa} = 0 \ (\kappa = 1, 2 \ldots n);$$ so that the ρ equations $\psi_i = 0$ are equivalent to $\rho - n$ [equations].—I hope to be able to settle in an analogous fashion the general case, where the scalars z_α are replaced by tensors $g_{\mu\nu}$.

Your most devoted,

Emmy Noether.

---

[2] See comments on this letter, *supra*, Chap. 1, p. 46.
[3] The text should read "independent of x_κ."

Appendix II
Letter from Emmy Noether to Felix Klein, 12 March 1918

Letter from Emmy Noether to Felix Klein, 12 March 1918, recto
(Niedersächsische Staats- und Universitätsbibliothek Göttingen)

Letter from Emmy Noether to Felix Klein, 12 March 1918, verso
(Nierdersächsische Staats- und Universitätsbibliothek Göttingen)

Transcription of the letter from Emmy Noether to Felix Klein
12 March 1918[1]

Erlangen, 12/3. 18

Sehr verehrter Herr Geheimrat!

Ich danke Ihnen sehr für die Zusendung Ihrer Notiz, und für Ihre Mitteilung über Runge's Ei des Kr. Columbus. Restlos bin ich nicht damit einverstanden; es versagt nämlich gerade im allereinfachsten Fall, dem der Homogenität erster Ordnung; allgemeiner, wenn in einem invarianten Variationsproblem $\delta \int \cdots \int f(z, \frac{\partial z}{\partial x} \cdots) dx_1 \ldots dx_n$ die z Skalare sind. Hier ist die Bedingung : $\sum \psi_i \frac{\partial z_i}{\partial x_\sigma} = 0 \ (\sigma = 1, 2 \ldots n)$ *identisch* erfüllt; es ist gerade die Lie'sche Differentialgleichung, und man kann auf keine Weise einen Energiesatz zwingen, es sei denn, man *postuliert* ihn an Stelle der Runge'schen Bedingung.

Ein Beispiel bietet das „Prinzip der kleinsten Wirkung"
$$\delta \int \sqrt{\left(\frac{dz_1}{dt}\right)^2 + \cdots \left(\frac{dz_n}{dt}\right)^2} \, dt = 0$$; hier tritt an Stelle des Energiesatzes die Abhängigkeit zwischen den Lagrange'schen Gleichungen: $\sum \psi_i \frac{dz_i}{dt} = 0$; [entsprechend der Invarianz gegenüber $t = \varphi(\bar{t})$, z Skalar] man kommt erst auf die gewöhnlichen Gleichungen $\frac{d^2 z_i}{dt^2} = 0$ zurück, wenn man den Energiesatz postuliert als nicht-invariante Zusatzbedingung zur eindeutigen Festlegung der z. – Baut man nun die Mechanik so auf, analog der Gravitationstheorie, daß man das „invariante" Prinzip der kleinsten Wirkung an die Spitze stellt, und nur *das* Mechanik nennt, was daraus folgt, also invariant ist, so besitzt diese Mechanik keinen Energiesatz, wohl aber eine Abhängigkeit zwischen den Gleichungen; und alles, worüber man sich jetzt wundert, ist schon dagewesen.

Diese Postulierung des Energiesatzes macht man auch immer beim Problem der geodätischen Linie $\delta \int \left(\frac{ds}{dt}\right) dt = 0$ – wovon ja das vorige ein Spezialfall ist – wenn man die Bogenlänge als Parameter einführt; d.h. $\frac{ds}{dt} = $ const., oder $\frac{d}{dt}\left(\frac{ds}{dt}\right) = 0$.

Bei meinen weiteren Untersuchungen habe ich jetzt gesehen, daß der Energiesatz versagt bei Invarianz gegenüber *jeder durch induzierte Transformationen*

---

[1] Manuscript in der Niedersächsische Staats- und Universitätsbibliothek, Göttingen, reproduced on the preceding pages. Publication kindly authorized by the Bibliothek.

der z erzeugten erweiterten Gruppe. Unter der allgemeinsten induzierten Transformation verstehe ich dabei, daß die z_i ersetzt werden durch Funktionen: $v_i(y) = \varphi_i\left(z, \frac{dz}{dx}, \ldots, \frac{d^\rho z}{dx^\rho}, x(y), \frac{dx}{dy}, \ldots, \frac{d^\sigma x}{dy^\sigma}\right)$, mit der einzigen Bedingung, daß der Identität in x auch die identische Transformation der z entspricht. Die Gruppe besteht dann aus der Substitution φ und ihren Wiederholungen; die Tensor-Substitution und entsprechende sind dadurch ausgezeichnet, daß die Wiederholungen von φ *formal mit* φ *identisch* werden, nur geschrieben in andern Variabeln; und deshalb hat man wohl diese allgemeineren Gruppen ganz übersehen. Hier muß auch die Umkehrung gelten, daß aus dem Versagen der Energiesätze Invarianz gegenüber dieser Gruppe folgt; doch bin ich mir über die Lie'schen Differentialgleichungen noch nicht ganz klar.

Die Feier am 5. ist sehr schön verlaufen, und meinem Vater sehr gut bekommen.

Mit besten Grüßen,

Ihre sehr ergebene

Emmy Noether.

Translation of the letter of Emmy Noether to Felix Klein, 12 March 1918[2]

Erlangen, 12 March 1918

Revered Privy Councilor,[3]

I thank you very much for sending me your note and your paper on Runge's Chr[istopher] Columbus's egg.[4] I absolutely cannot agree with that; it is actually not even valid in the simplest case, that of homogeneity of the first order, nor more generally, when in an invariant variational problem $\delta \int \cdots \int f(z, \frac{\partial z}{\partial x} \cdots) dx_1 \ldots dx_n$ the [dependent variables] z are scalars.[5] Here the condition $\sum \psi_i \frac{\partial z_i}{\partial x_\sigma} = 0 \ (\sigma = 1, 2 \ldots n)$ is satisfied *identically*; it is precisely Lie's differential equation, and one cannot, in any manner, obtain [the validity of] an energy law, unless one *assumes* it instead of Runge's condition.

An example is provided by the "principle of least action," $\delta \int \sqrt{\left(\frac{dz_1}{dt}\right)^2 + \cdots \left(\frac{dz_n}{dt}\right)^2} \, dt = 0$; here there appears, instead of the energy law, the

---

[2] See comments on this letter, *supra*, Chap. 1, p. 41 and p. 47.

[3] The honorary title of Privy Councilor had been granted by the emperor to several of the most distinguished professors.

[4] The expression "Christopher Columbus's egg" used to be used, and is still used, to describe an unexpected, simple, but abrupt solution to a problem. In his letter to Einstein of 20 March 1918, cited, *supra*, Chap. 1, p. 42, note 56, Klein used that expression to describe the contents of Carl Runge's lecture of 8 March. Runge had proposed a particular choice of coordinates in which it was possible to show that the law of conservation of energy amounted to the vanishing of a divergence.

[5] I.e., remain unchanged. Cf., Noether [1918c], section 1.

identity[6] among [the left-hand sides of] the Lagrange equations, $\sum \psi_i \dfrac{dz_i}{dt} = 0$; (corresponding to the invariance under [the transformations] $t = \varphi(\bar{t})$, z scalar[7]); one only recovers the usual equations $\dfrac{d^2 z_i}{dt^2} = 0$ when one assumes the energy law as a noninvariant additional condition in order to determine z uniquely.—If one then constructs mechanics in this way, in a fashion analogous to the theory of gravitation, by supposing at the outset the "invariant" principle of least action, and if one calls "the" mechanics only what one deduces from it, and is consequently invariant, then this mechanics does not possess any energy law, but, instead, an identity among the [Lagrange] equations;[8] and everything that astonished us up to now was already present.

Also, one always supposes the energy law in the problem of geodesic curves $\delta \displaystyle\int \left(\dfrac{ds}{dt} \right) dt = 0$—of which the preceding is indeed a particular case—if one introduces the line element as a parameter; that is to say $\dfrac{ds}{dt} = \text{const.}$, or $\dfrac{d}{dt} \left(\dfrac{ds}{dt} \right) = 0$.

As a result of my further research, I have now seen that the energy law is not valid[9] in the case of invariance under every *extended group generated by the transformation induced by the z.* By the term, the most general induced transformation, I mean that the z_i are replaced by functions $v_i(y) = \varphi_i\left(z, \dfrac{dz}{dx}, \ldots, \dfrac{d^p z}{dx^p}, x(y), \dfrac{dx}{dy}, \ldots, \dfrac{d^\sigma x}{dy^\sigma}\right)$, subject to the unique condition that the identity on x corresponds to the identical transformation on z. The group is constituted by the substitution φ and its successive compositions; the tensorial substitution and those which correspond [to it] are characterized by the fact that the compositions of φ *become formally identical to* φ, just written with other variables; and that is why these more general groups have been completely ignored. Here as well the converse should be valid, because the lack of an energy law implies the invariance under this group;[10] but in regard to Lie's differential equations, this is not yet entirely clear to me.

The festivities of the 5th were quite successful and did my father much good.

With best wishes,

Your most devoted,

Emmy Noether

---

[6] We translate *Abhängigkeit* by "identity," as we have done in the translation of Noether's article.

[7] I.e., z being unchanged.

[8] This is a preliminary formulation of Noether's fundamental result, which would be published in Section 6 of her article. Noether here rectifies Hilbert's affirmation, which had only applied to general relativity.

[9] Cf., Noether [1918c], section 6, p. 20, in the above translation. See, *supra*, Chap. 1, p. 47.

[10] The converse will be stated more precisely in sections 4 and 6 of her article.

Appendix III
Letter from Felix Klein to Wolfgang Pauli, 8 March 1921

Translation of the letter printed in Pauli [1979], pp. 27–28 [1]

Göttingen, 8 March 1921

Dear Mr. Pauli,

Your mail reached me fine and while thanking you and Sommerfeld, I must say that I am also in agreement with this way of proceeding. I also wrote to Teubner just now [...].

My own work contains no speculations of natural philosophy, but only a straightening out of the process of math[ematical] thinking which often seemed to me, in Laue's writings for one example, quite tortured and thus impenetrable. I report with pleasure what Einstein writes concerning my third note[2]: he feels quite happy, like a child whose mother has offered him a chocolate bar (in his personal comments Einstein is always this charming, in total contradiction with the wild publicity that surrounds him). Another point that I have tried to clarify in my course notes is the historical state of affairs. It is true that Poincaré's first note[3] in the *Comptes Rendus* 140 precedes Einstein['s paper],[4] and that it is Poincaré who has shown for the first time (in the *Rendiconti di Palermo*[5]) that in Lorentz there was a *group* of transformations. Whence a contradiction which alone can explain why P[oincaré] 1911, in his lecture at Göttingen "Sur la nouvelle mécanique"[6] does not even mention the

---

[1] Translation published with permission of Springer, Berlin. See comments on this letter, *supra*, Chap. 4, p. 93.

[2] Klein [1918c].

[3] H. Poincaré, Sur la dynamique de l'électron, *Comptes rendus hebdomadaires des séances de l'Académie des sciences*, 140 (1905), pp. 1504–1508.

[4] A. Einstein, Zur Elektrodynamik bewegter Körper, *Annalen der Physik*, 17 (1905), pp. 891–921.

[5] H. Poincaré, Sur la dynamique de l'électron, *Rendiconti del Circolo Matematico di Palermo*, 21 (1906), pp. 129–175.

[6] "On the new mechanics" (in French in the original).

name of Einstein. I would consider important that these facts and other, similar ones appear in your report.* After all, enough remains on that account for Einstein.

You may in any event keep my notes until the complete work for the *Enzy[klopädie]* article is finished. It may be appropriate to refer to the Dutchmen even more strongly than I did, regarding the details. I was hindered for a long time by the language, with which I am not familiar, because I have only the originals written in Dutch at hand.

Another much simpler matter. After Batemann [*sic*] had observed[2] that Maxwell's equations are transformed into themselves under his G_{15}, it is clear from E. Noether's theorems that there are 15 divergence relations for those equations. In the meantime, one of my students, Dr. Bessel-Hagen, has written them explicitly and he has found a few of them that were apparently, until now, unknown in the literature. But I would prefer, before accepting this for the *Mathemat[ischen] Annalen*,[3] that they be checked on the physics side. Dr. Bessel-Hagen has just left for a vacation in Berlin (address: Kurfürstendamm 200, Berlin W.) and I advised him, because he has some personal connections, to show the matter to Planck. It may also be very useful if he gets in touch with you. I would also like to ask you if you would be able to say something about this question, and if I can possibly suggest to Bessel-Hagen to get in touch with you.

Unfortunately, I cannot myself get involved more deeply in these questions. I must now prepare the vol. II of my complete works and consequently I am now deeply immersed once more in the theory of algebraic equations.

Yours truly,

Klein

---

* The Einstein–Hilbert relation also belongs in this context, and for that reason in vol. I of my *Works*, I gave the exact dates. E[instein] and H[ilbert] have met and corresponded on several occasions, but they did not use the same language, which is not rare in the case of mathematicians who are working simultaneously. But physicists maintain a deathly silence about Hilbert's achievement which is, admittedly, expressed in a very awkward presentation.

[2] The conformal invariance of Maxwell's equations was the subject of an article by Bateman published in 1910. See, *supra*, Chap. 4, p. 91, note 2, and Rowe [1999], p. 211.

[3] Bessel-Hagen's article was submitted on 3 March 1921.

Appendix IV
Letter from Emmy Noether to Albert Einstein,
7 January 1926

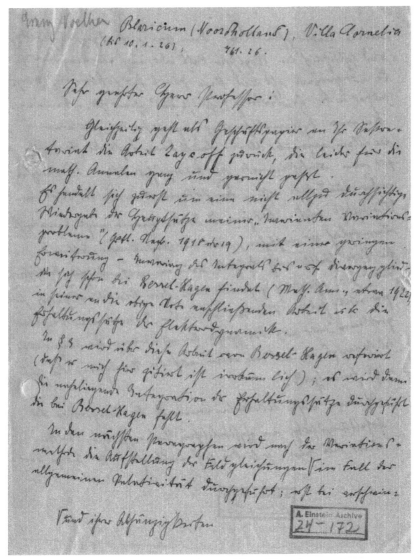

Letter from Emmy Noether to Albert Einstein, page 1
(Einstein Archive, Jerusalem, reproduced with permission of the Archive)

Letter from Emmy Noether to Albert Einstein, page 2
(Einstein Archive, Jerusalem, reproduced with permission of the Archive)

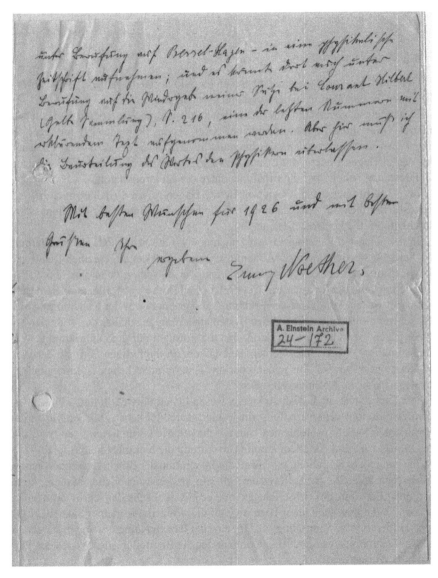

Letter from Emmy Noether to Albert Einstein, page 3
(Einstein Archive, Jerusalem, reproduced with permission of the Archive)

Transcription of the letter from Emmy Noether to Albert Einstein
7 January 1926[1]

Blaricum (Noordholland), Villa Cornelia, 7.Jan. 26
(bis 10.Jan. 26)

Sehr geehrter Herr Professor:

Gleichzeitig geht als Geschäftspapier an Ihr Sekretariat die Arbeit Zaycoff zurück, die leider für die math. Annalen ganz und garnicht passt.

Es handelt sich zuerst um eine nicht allzu durchsichtige Wiedergabe der Hauptsätze meiner "Invarianten Variationsprobleme" (Gött. Nachr. 1918 oder 19), mit einer geringen Erweiterung–Invarianz des Integrals bis auf Divergenzglied–die sich schon bei Bessel-Hagen findet (Math. Ann., etwa 1922), in seiner an die obige Note anschliessenden Arbeit über die Erhaltungssätze der Elektrodynamik.

In § 3 wird über diese Arbeit von Bessel-Hagen referiert (daß er mich hier zitiert ist irrtümlich); es wird dann die naheliegende Integration der Erhaltungssätze durchgeführt, die bei Bessel-Hagen fehlt.

In den nächsten Paragraphen wird nach der Variationsmethode die Aufstellung der Feldgleichungen und ihrer Abhängigkeiten im Fall der allgemeinen Relativität durchgeführt; erst bei verschwindenden elektrischen Vektor, denn ohne diese Spezialisierung und schliesslich im Weyl'schen Falle oder noch allgemeiner; da nur gerechnet ist und kein Wort der Erklärung gesagt (ausser in der Einleitung), so ist das schwer zu erkennen. Die ganze Systematisierung gegenüber den früheren–vor allem gegenüber Klein–beruht darin, daß die Formeln dür eine unbestimmte Wirkungsfunktion W berechnet werden, and daß erst in die fertigen Formeln der Wert von W eingesetzt wird. Für jemand der die Theorie nicht kennt, ist unmöglich zu verstehen, was die Rechnungen sollen.

Die Sache stellt auch deshalb keinen wesentlichen Fortschritt dar, weil schliesslich fast alle mit dem Variationsprinzip hier gearbeitet haben. Mir kam es in den "Invarianten Variationsproblemen" nur auf die scharfe Formulierung der Tragweite des Prinzips an, und vor allem auf die Umkehrung die hier nicht herein spielt.

Ich kann nicht beurteilen, inwieweit die Integration der Erhaltungssätze von physikalischem Interesse ist. Sollte es der Fall sein, so liesse sich vielleicht dieser kurze Teil unter Berufung auf Bessel-Hagen–in eine physikalische Zeitschrift aufnehmen; und es könnte dort auch unter Berufung auf die Wiedergabe meiner Sätze bei Courant–Hilbert (Gelbe Sammlung), S.216, eine der letzten Nummern, mit erklärendem Text aufgenommen werden. Aber hier muss ich die Beurteilung des Wertes den Physikern überlassen.

Mit besten Wünschen für 1926 und mit besten Grüßen

Ihre ergebene

Emmy Noether.

---

[1] Manuscript 24-172 of the Einstein Archive, Jerusalem. Publication kindly authorized by the Archive. I thank Prof. Peter Roquette for communicating a transcript of this document to me.

Translation of the letter from Emmy Noether to Albert Einstein, 7 January 1926[2]

Blaricum (North Holland), Villa Cornelia, 7 January 1926
(until 10 January 1926)

Dear Professor,

Presently I am returning in the mail to your secretariat as commercial paper the article by Zaycoff, which unfortunately is by no means suitable for the Math[ematische] Annalen.

It is first of all a restatement that is not at all clear of the principal theorems of my "Invariante Variationsprobleme" (Gött[inger] Nachr[ichten], 1918 or 19), with a slight generalization—the invariance of the integral up to a divergence term—which can actually already be found in Bessel-Hagen (Math[ematische] Ann[alen], around 1922), in his work on the conservation laws of electrodynamics which is related to the above-mentioned note.

In section 3 there is a reference to Bessel-Hagen's work (citing me there is an error); then the author performs the obvious integration of the conservation laws which is not in Bessel-Hagen.

In the following paragraphs he establishes by the method of the calculus of variations the field equations and their identities in the case of general relativity; first in the case where the electric field vanishes, then without that assumption, and finally in Weyl's case or in a still more general case; since there are only calculations and since there is not a single word of explanation (except in the introduction), this is hard to understand. All the systematization with respect to earlier work—above all with respect to Klein—depends on the fact that the formulae are established for any action functional W and that the value of W is only specified in the final formulae. It would be impossible for someone who does not know the theory to understand the calculations.

Therefore, the article does not represent any real progress because, in conclusion, nearly everyone at this point has worked with the variational principle. For me, what was the most important in the "Invariante Variationsprobleme" was to state in a rigorous fashion the significance of the principle and, above all, to state the converse, which does not appear here.

I cannot appreciate to what extent the integration of the conservation laws is interesting from the point of view of physics. If that were the case, it might be possible to induce a physics journal to accept this limited part, with a reference to Bessel-Hagen; it would also be possible to introduce in it a reference to the statement of my theorems in Courant–Hilbert (Yellow Collection), p. 216, one of the most recent volumes, with an explanatory text. But for this, I must leave it to the physicists to judge the value.

With my best wishes for 1926 and my best regards,

Your devoted,

Emmy Noether.

---

[2] See comments on this letter, *supra*, Chap. 1, p. 51.

Appendix V
Lectures delivered at the Mathematical Society of Göttingen, 1915–1919

When one consults the second part of volumes 24 to 28 of the *Jahresbericht der Deutschen Mathematiker-Vereinigung* for the years 1915–1919, it appears that Noether, referred to as either "Emmy Noether" or as "Frl. Noether," had frequently delivered papers before the *Mathematische Gesellschaft zu Göttingen*. Here is a list of the papers that she delivered and of the papers delivered by colleagues that were related to her research interests.

29 June 1915, Einstein, *Über Gravitation* (On gravitation), vol. 24, Part 2, p. 68.

13 July 1915, Noether, *Endlichkeitsfragen der Invariantentheorie* (Questions of finiteness in the theory of invariants), vol. 24, Part 2, p. 68.

9 November 1915, Noether, *Über ganze transzendente Zahlen* (On integral transcendental numbers), vol. 24, Part 2, p. 111.

The next week, 16 November 1915, Hilbert lectured on the *Grundgleichungen der Physik* (Fundamental equations of physics), vol. 24, Part 2, p. 111.

30 November and 7 December 1915, Hilbert and Carathéodory lectured about the theory of invariants, *Über Invariantentheorie*, vol. 24, Part 2, p. 111.

25 January 1916, Hilbert, *Invariantentheorie und allgemeiner Energiesatz* (Invariant theory and the generalized law of energy), vol. 25, Part 2, p. 31.

1 February 1916, Noether, *Alternative bei nichtlinearen Gleichungsystemen* (Alternatives in systems of nonlinear equations), vol. 25, Part 2, p. 31.

23 May 1916, Noether, *Gleichungen mit vorgeschriebener Gruppe* (Equations with prescribed [Galois] group), vol. 25, Part 2, p. 66.

23 January 1917, Carathéodory, *Variationsprobleme mit symmetrischer Transversalität* (Variational problems with symmetrical transversality); Hilbert, *Nichteuklidische Geometrie und die neue Gravitationstheorie* (Noneuclidean geometry and the new gravitation theory), vol. 25, Part 2, p. 113.

19 June 1917, Noether, *Laskers Zerlegungssatz der Modultheorie* (Lasker's decomposition theorem in the theory of modules), vol. 26, Part 2, p. 31.

4 December 1917, Klein, *Über Herglotz, Interpretation des Hilbertschen Krümmung und des zugehörigen Gravitationstensor* (On Herglotz, Interpretation of Hilbert's curvature and the associated gravitation tensor), vol. 26, Part 2, p. 70.

15 January 1918, Noether, *Über Invarianten beliebiger Differentialausdrücke* (On the invariants of arbitrary differential expressions). The announcement of the lecture contains an abstract, vol. 27, Part 2, p. 28.

22 January 1918, Klein lectured on Hilbert's first note on the foundations of physics (Hilbert [1915]), and Hilbert continued the discussion 29 January.[1] The announcement of Hilbert's lecture contains an abstract that states the result:

> The "conservation laws" of classical mechanics that apply to the mechanics of continua (the energy-momentum theorems) are already contained in the field equations in the new theory initiated by Einstein; they therefore lose their independent importance.[2]

This lecture may have been the immediate cause of Noether's efforts to clarify the nature of conservation laws in general relativity.

7 May 1918, Klein discussed Einstein's ideas of 1917 on cosmology and continued on June 11. Both announcements contain extended abstracts, vol. 27, Part 2, pp. 42–43 and p. 44.

14–17 May 1918, Max Planck was invited to give a series of four lectures on the present state of the quantum theory, vol. 27, Part 2, p. 43.

3 June 1918, Runge reported on Einstein's communication on gravitational waves of 31 January 1918 to the Berlin Academy. The announcement of the lecture contains an abstract, vol. 27, Part 2, pp. 43–44.

4 June 1918, Hilbert, *Energiesatz für die Bewegung eines Planeten in der neuen Gravitationstheorie* (Energy theorem for the movement of a planet in the new gravitation theory). The announcement of the lecture contains a short abstract, vol. 27, Part 2, p. 44.

4 July 1918, Klein lectured about Einstein's new communication of 16 May 1918 to the Berlin Academy on the energy theorem in general relativity. The announcement of the lecture contains an abstract, vol. 27, Part 2, p. 45.

15 July 1918, Hilbert lectured on Weyl's paper delivered 2 May in Berlin on *Gravitation und Elektrizität* (Gravitation and electricity). The announcement of the lecture contains a short abstract mentioning the introduction of an arbitrary multiplicative factor λ, vol. 27, Part 2, p. 46.

---

[1] See Rowe [1999], p. 212.

[2] "Die für die Mechanik der Kontinua geltenden 'Erhaltungssätze' der klassischen Mechanik (die Impuls-Energie-Sätze) sind bei der neuen, von Einstein inaugurierten Theorie in den Feldgleichungen bereits mit enthalten; sie verlieren damit ihre selbständige Bedeutung" (vol. 27, Part 2, p. 28).

22 July 1918, Klein lectured on Hilbert's energy vector. The announcement of the lecture contains an abstract, vol. 27, Part 2, pp. 46–47.

23 July 1918, Noether, *Invariante Variationsprobleme* (Invariant variational problems). The announcement of the lecture contains the following abstract.

In connection with the research on Hilbert's energy vector, the lecturer stated the following general theorems:

If the first variation of an integral is invariant under a finite continuous group with ρ essential parameters, then ρ linear combinations of the Lagrangian derivatives of the integral become divergences. In particular, one thus obtains in the one-dimensional case, where the divergences become total differentials, ρ first integrals of the differential equations given by the vanishing of the first variation.

If the first variation is invariant under an infinite continuous group with ρ arbitrary functions, then there exist among the Lagrangian derivatives and their differentials ρ linear identities, so that ρ equations become a consequence of the others.

For both theorems the converse holds.[3]

5 November 1918, Noether, *Endlichkeit ganzzahliger binären Invarianten* (Finiteness of integral binary invariants). The announcement of the lecture contains an abstract, vol. 28, Part 2, p. 29.

26 November 1918, Noether, *Über ganzzahlige Polynome und Potenzreihen* (On integral polynomials and power series). The announcement of the lecture contains an abstract, vol. 28, Part 2, pp. 29–30.

A year and a half later, Weyl came to lecture (11 May 1920) on *Das Kontinuum* (The continuum), but Noether's lectures of that year did not deal with problems related to physics but rather the theory of modules and the arithmetic theory of algebraic functions. Later in the year, at the meeting of the German Mathematical Society in September, while Weitzenböck lectured on the theory of invariants in the new physics, Noether spoke not on invariants and physics but on questions in the theory of modules and ideals.

---

[3] "In Zusammenhang mit den Untersuchungen über den Hilbertschen Energievektor hat die Referentin folgende allgemeine Sätze aufgestellt:

Gestattet die erste Variation eines Integrals eine endliche kontiuierliche Gruppe von ρ wesentlichen Parametern, so werden ρ lineare Verbindungen der Lagrangeschen Ableitungen des Integrals zu Divergenzen. Insbesondere kennt man also im eindimensionalen Falle, wo die Divergenzen zu totalen Differentialquotienten werden, ρ erste integrale der durch Nullsetzen der ersten Variation gegebenen Differentialgleichungen.

Gestattet die erste Variation eine unendliche kontiuierliche Gruppe mit ρ willkürlichen Funktionen, so bestehen zwischen den Lagrangeschen Ableitungen und ihren Differentialquotienten ρ lineare Beziehungen, so daß ρ Gleichungen eine Folge der übrigen werden.

Zu beiden Sätzen gilt die Umkehrung" (vol. 27, Part 2, p. 47).

References

For short, we shall cite the *Nachrichten von der [Königlichen] Gesellschaft der Wissenschaften zu Göttingen, Mathematisch-physikalische Klasse* as *Göttinger Nachrichten*, Noether's *Gesammelte Abhandlungen / Collected Papers* as *Abhandlungen*, and *The Collected Papers of Albert Einstein* as *Collected Papers*.

ABRAHAM (Ralph), MARSDEN (Jerrold E.)
[1967] *Foundations of Mechanics*, Reading, MA: Benjamin/Cummings Publishing Co., 1967; 2nd ed., revised, enlarged and reset, 1978.

ALEXANDROV (Pavel S.)
[1936] In memory of Emmy Noether [in Russian], *Uspekhi Matematicheskikh Nauk*, 2 (1936), pp. 254–265; English translations, in Brewer and Smith [1981], pp. 99–111, and in Noether, *Abhandlungen*, pp. 1–11.
[1979] Pages from an autobiography [in Russian], *Uspekhi Matematicheskikh Nauk*, 34 (6) (1979), pp. 219–249; English translation, *Russian Mathematical Surveys*, 35 (3) (1980), pp. 315–359.

AMES (Joseph S.), MURNAGHAN (Francis D.)
[1929] *Theoretical Mechanics: An Introduction to Mathematical Physics, s.l.*: Ginn and Company, 1929; reprint, New York: Dover, 1958.

ANDERSON (Dan)
[1973] Noether's theorem in generalized mechanics, *Journal of Physics A*, 6 (1973), pp. 299–305.

ANDERSON (Ian M.)
[1992] Introduction to the variational bicomplex, in *Mathematical Aspects of Classical Field Theory* (Seattle 1991), Mark J. Gotay, Jerrold E. Marsden and Vincent Moncrief, eds., Contemporary Mathematics, vol. 132, Providence, RI: American Mathematical Society, 1992, pp. 51–73.

ANDERSON (Ian M.), DUCHAMP (Thomas Eugene)
[1980] On the existence of global variational principles, *American Journal of Mathematics*, 102 (1980), pp. 781–868.

ANDERSON (Ian M.), POHJANPELTO (Juha)
[1994] Variational principles for differential equations with symmetries and conservation laws, I, *Mathematische Annalen*, 299 (1994), pp. 191–222.

ANDERSON (James L.)
  [1967] *Principles of Relativity Physics*, New York: Academic Press, 1967.

ANDERSON (James L.), BERGMANN (Peter Gabriel)
  [1951] Constraints in covariant field theories, *Physical Review*, 83 (1951), pp. 1018–1025.

ANDERSON (Robert L.), IBRAGIMOV (Nail H.)
  [1979] *Lie–Bäcklund Transformations in Applications*, SIAM Studies in Applied Mathematics, Philadelphia: SIAM, 1979.

ANDERSON (Robert L.), KUMEI (Sukeyuki), WULFMAN (Carl E.)
  [1972] Generalization of the concept of invariance of differential equations, Results of applications to some Schrödinger equations, *Physical Review Letters*, 28 (1972), pp. 988–991.

ARNOLD (Vladimir I.)
  [1974] *Matematicheskie metody klassicheskoj mekhaniki* [Russian], Moscow: Nauka, 1974; 2nd ed., 1979; French translation by Djilali Embarek, *Les Méthodes mathématiques de la mécanique classique*, Moscow: Mir, 1976; English translation by K. Vogtman and Alan Weinstein, *Mathematical Methods of Classical Mechanics*, Graduate Texts in Mathematics, vol. 60, New York, Heidelberg, Berlin: Springer-Verlag, 1978; 2nd English ed., 1989; German translation of the 2nd Russian ed., *Mathematische Methoden der klassischen Mechanik*, Basel: Birkhäuser, 1988.

AUTONNE (Léon)
  [1891] Sur une application des groupes de M. Lie, *Comptes rendus de l'Académie des sciences de Paris*, 112 (1891), pp. 570–573.

BARGMANN (Valentine)
  [1960] Relativity, in *Theoretical Physics in the Twentieth Century. A Memorial Volume to Wolfgang Pauli*, Markus Fierz and Victor F. Weisskopf, eds., New York, London: Interscience, 1960, pp. 187–198.

BARINAGA (J.)
  [1935] Emmy Nöther, *Revista Matematica Hispano-Americana* (2), 10 (1935), pp. 162–163.

BARUT (Asim Orhan)
  [1964] *Electrodynamics and Classical Theory of Fields and Particles*, New York: Macmillan; London: Collier Macmillan, 1964; corrected ed., New York: Dover, 1980.

BATEMAN (Harry)
  [1910] The transformation of the electrodynamical equations, *Proceedings of the London Mathematical Society*, 2nd series, 8 (1910), pp. 223–264.

BELINFANTE (Frederik Jozef)
  [1939] On the spin angular momentum of mesons, *Physica*, 6, no. 9 (1939), pp. 887–898.
  [1940] On the current and the density of the electric charge, the energy, the linear momentum and the angular momentum of arbitrary fields, *Physica*, 7, no. 5 (1940), pp. 449–474.

BENYOUNÈS (Michèle)
  [1987] Actions des symétries généralisées sur les lois de conservation des systèmes lagrangiens, Thèse de troisième cycle, Université des Sciences et Techniques de Lille Flandres Artois, July 1987 (mimeographed).

BERGMANN (Birgit), EPPLE (Moritz), UNGAR (Ruti), eds.,
  [2012] *Transcending Tradition, Jewish Mathematicians in German-Speaking Academic Culture*, Berlin, Heidelberg: Springer Verlag, 2012.

BERGMANN (Peter Gabriel)
  [1942] *Introduction to the Theory of Relativity*, preface by Albert Einstein, Englewood Cliffs, NJ: Prentice-Hall, 1942; reprints, 1942, 1946, 1947, 1948, 1950; 2nd ed., revised, with two appendices, New York: Dover, 1976.
  [1949] Nonlinear field theories, *Physical Review*, 75 (1949), pp. 680–685.
  [1958] Conservation laws in general relativity as the generators of coordinate transformations, *Physical Review*, 112 (1958), pp. 287–289.

BERGMANN (Peter Gabriel), THOMSON (Robb)
[1953] Spin and angular momentum in general relativity, *Physical Review*, 89 (1953), pp. 400–407.

BESSEL-HAGEN (Erich)
[1921] Über die Erhaltungssätze der Elektrodynamik, *Mathematische Annalen*, 84 (1921), pp. 258–276.

BLAKER (J. Warren), TAVEL (Morton A.)
[1974] The applications of Noether's theorem to optical systems, *American Journal of Physics*, 42 (1974), pp. 857–861.

BLISS (Gilbert A.),
[1920] Some recent developments in the calculus of variations, *Bulletin of the American Mathematical Society*, 26 (1920), pp. 343–361.
[1925] *Calculus of Variations*, Chicago, IL: Mathematical Association of America, Open Court Publishing Company, 1925.

BLOCH (Anthony M.), KRISHNAPRASAD (P. S.), MARSDEN (Jerrold E.), MURRAY (Richard M.)
[1996] Nonholonomic mechanical systems with symmetry, *Archive for Rational Mechanics and Analysis*, 136 (1996), pp. 21–99.

BLUMAN (George W.), KUMEI (Sukeyuki)
[1989] *Symmetries and Differential Equations*, New York, Berlin, Heidelberg: Springer-Verlag, 1989.

BOGOLYUBOV (Nikolai N.), CHIRKOV (Dmitrii V.)
[1957] *Vvedenie v teoriju kvantovannykh polej* [Russian], Moscow: Gosudarstvennoe Izdateltsvo Tekhniko-teoreticheskoj Literatury, 1957; English edition revised by the authors, translated by G. M. Volkoff, *Introduction to the Theory of Quantized Fields*, New York: Wiley, 1959; French translation by A. Bloch, *Introduction à la théorie quantique des champs*, Paris: Dunod, 1960.

BOHR (Niels), KRAMERS (Hendrik), SLATER (John C.),
[1924] The quantum theory of radiation, *Philosophical Magazine*, 47 (1924), pp. 795–802; German version in *Zeitschrift für Physik*, 24 (1924), pp. 69–87.

BOYER (Timothy H.)
[1966] Derivation of conserved quantities from symmetries of the Lagrangian in field theory, *American Journal of Physics*, 34 (1966), pp. 475–478.
[1967] Continuous symmetries and conserved currents, *Annals of Physics*, 42 (1967), pp. 445–466.

BRADING (Katherine A.)
[2005] A note on general relativity, energy conservation, and Noether's theorems, in *The Universe of General Relativity* (Proceedings of the Conference on the History of General Relativity, Amsterdam 2002), Jean Eisenstaedt and Anne J. Kox, eds., Einstein Studies, vol. 11, Boston: Birkhäuser, 2005, pp. 125–135.

BRADING (Katherine A.), BROWN (Harvey R.)
[2003] Symmetries and Noether's theorems, in Brading and Castellani [2003], pp. 89–109.

BRADING (Katherine A.), CASTELLANI (Elena), eds.
[2003] *Symmetries in Physics, Philosophical Reflections*, Cambridge: Cambridge University Press, 2003.

BREWER (James W.), SMITH (Martha K.), eds.
[1981] *Emmy Noether, A Tribute to Her Life and Work*, Monographs and Textbooks in Pure and Applied Mathematics, vol. 69, New York, Basel: Marcel Dekker, 1981.

BYERS (Nina)

[1996] The life and times of Emmy Noether, Contributions of Emmy Noether to particle phyiscs, in *History of Original Ideas and Basic Discoveries in Particle Physics*, H. B. Newman and T. Ypsilantis, eds., New York: Plenum Press, 1996, pp. 945–964.

[1999] E. Noether's discovery of the deep connection between symmetries and conservation laws, in Teicher [1999], pp. 67–81.

[2006] Symmetries and conservation laws in physics including energy conservation in relativity theory; modern algebra, in *Out of the Shadows. Contributons of Twentieth-Century Women to Physics*, Nina Byers and Gary Williams, eds., pp. 83–96.

CANDOTTI (E.), PALMIERI (C.), VITALE (Bruno)

[1970] On the inversion of Noether's theorem in the Lagrangian formalism, II. Classical field theory, *Il Nuovo Cimento*, 70 A (1970), pp. 233–279.

CARATHÉODORY (Constantin)

Gesammelte mathematische Schriften, vol. 1, Munich: C. H. Beck'sche Verlagsbuchhandlung, 1954.

[1929] Über die Variationsrechnung bei mehrfachen Integralen, *Acta litterarum ac scientiarum regiae universitatis Hungaricae Francisco-Josephinae, Sectio Scientiarum Mathematicarum*, Szeged, 4, 1929, pp. 193–216; *Gesammelte mathematische Schriften*, vol. 1, pp. 401–426.

[1935] *Variationsrechnung und partielle Differentialgleichungen erster Ordnung*, Berlin: Teubner, 1935; English translation by Robert B. Dean, *Calculus of Variations and Partial Differential Equations of the First Order*, 2nd ed., New York: Chelsea, 1982.

CARMELI (Moshe), LEIBOWITZ (Elhanan), NISSANI (Noah)

[1990] *Gravitation: SL(2,ℂ) Gauge Theory and Conservation Laws*, Singapore: World Scientific, 1990.

CARTAN (Élie)

[1896] Le principe de dualité et certaines intégrales multiples de l'espace tangentiel et de l'espace réglé, *Bulletin de la Société Mathématique de France*, 24 (1896), pp. 140–177.

[1922] *Leçons sur les invariants intégraux*, Paris: J. Hermann, 1922.

CARTIER (Pierre), DEWITT-MORETTE (Cécile)

[2006] *Functional Integration. Action and Symmetries*, Cambridge: Cambridge University Press, 2006.

CATTANEO (Alberto S.), FIORENZA (Domenico), LONGONI (Riccardo)

[2006] Graded Poisson algebras, in *Encyclopedia of Mathematical Physics*, Jean-Pierre Françoise, Gregory L. Naber and Tsou Sheung Tsun, eds., Amsterdam, etc.: Academic Press/Elsevier Science, 2006, vol. 2, pp. 560–567.

CATTANEO (Carlo)

[1966] Conservation laws, *Annales de l'Institut Henri Poincaré* (A) Physique théorique, 4 (1966), pp. 1–20.

[1969] Invarianza e conservazione [Italian with English abstract], *Rendiconti del Seminario Matematico e Fisico di Milano*, 39 (1969), pp. 196–222.

CATTANI (Carlo), DE MARIA (Michelangelo)

[1993] Conservation laws and gravitational waves in general relativity (1915–1918), in Earman, Janssen and Norton [1993], pp. 63–87.

CAYLEY (Arthur)

Collected Mathematical Papers, vol. 1, Cambridge: Cambridge University Press, 1889.

[1845] On the theory of linear transformations, *Cambridge Mathematical Journal*, 4 (1845), pp. 193–209; translated into French with additions as "Mémoire sur les hyperdéterminants," *Journal für die reine und angewandte Mathematik*, 30 (1846), pp. 1–37; *Collected Mathematical Papers*, vol. 1, pp. 80–94 and 117.

CHANDRASEKHARAN (Komaravolu), ed.
[1986] *Hermann Weyl 1885–1985, Centenary Lectures*, Berlin, Heildelberg, New York: Springer-Verlag for the Eidgenössische Technische Hochschule Zürich, 1986.

CHERNOFF (Paul R.), MARSDEN (Jerrold E.)
[1974] *Properties of Infinite Dimensional Hamiltonian Systems*, Lecture Notes in Mathematics, 425, Berlin, Heildelberg, New York: Springer-Verlag, 1974.

CHOQUET-BRUHAT (Yvonne)
[1984] Positive-energy theorems, in *Relativity, Groups and Topology*, II (Les Houches, 1983), Bryce S. DeWitt and Raymond Stora, eds., Amsterdam: North-Holland, 1984, pp. 739–785.

COLEMAN (Albert John)
[1997] Groups and physics—Dogmatic opinions of a senior citizen, *Notices of the American Mathematical Society*, 44 (1) (January 1997), pp. 8–17.

CORRY (Leo), RENN (Jürgen), STACHEL (John)
[1997] Belated decision in the Hilbert–Einstein priority dispute, *Science*, 278 (1997), pp. 1270–1273.

CORSON (Edward Michael)
[1953] *Introduction to Tensors, Spinors and Relativistic Wave Equations*, Glasgow: 1953; reprint, New York: Chelsea, *s.d.*.

COURANT (Richard), HILBERT (David)
[1924] *Methoden der mathematischen Physik*, vol. 1, Berlin, Heidelberg: J. Springer, 1924; 2nd ed., corrected, 1931; English translation, *Methods of Mathematical Physics*, New York: Wiley-Interscience, 1953, numerous reprints.

CUNNINGHAM (Ebenezer)
[1910] The principle of relativity in electrodynamics and an extension thereof, *Proceedings of the London Mathematical Society* (2) 8 (1910), pp. 77–98 (meeting of 11 February 1909).

CURTIS (Charles W.)
[1999] *Pioneers of Representation Theory: Frobenius, Burnside, Schur, and Brauer*, History of Mathematics, vol. 15, Providence, RI: American Mathematical Society; London: London Mathematical Society, 1999.

CUSHMAN (Richard H.), BATES (Larry M.)
[1997] *Global Aspects of Classical Integrable Systems*, Basel, Boston, Berlin: Birkhäuser, 1997.

DASS (Tulsi)
[1966] Conservation laws and symmetries, *Physical Review* (2), 150 (1966), pp. 1251–1255.

DEDECKER (Paul)
[1975] On the generalization of symplectic geometry to multiple integrals in the calculus of variations, in *Differential Geometrical Methods in Mathematical Physics* (Bonn, 1975), K. Bleuler and A. Reetz, eds., Lecture Notes in Mathematics, 570, Berlin: Springer-Verlag, 1977, pp. 395–456.

DE DONDER (Théophile)
[1912] Sur les invariants du calcul des variations, *Comptes rendus de l'Académie des sciences de Paris*, 155 (1912), pp. 577–580 and 1003–1005.
[1916] Les équations différentielles du champ gravifique d'Einstein créé par un champ électromagnétique de Maxwell–Lorentz, *Koninklijke Akademie van Wetenschappen te Amsterdam, Verslagen van de Gewone Vergaderingen der Wis- en Natuurkundige Afdeeling*, 25 (1916–1917), pp. 153–156 (27 May 1916).

[1929] Théorie invariantive du calcul des variations (première communication), *Bulletins de l'Académie Royale de Belgique*, 15 (1929), pp. 7–25.

[1935] *Théorie invariantive du calcul des variations*, Paris: Gauthier-Villars et Cie, 1935 (first edition, 1931).

DELIGNE (Pierre), FREED (Daniel S.)
[1999] Classical field theory, in *Quantum Fields and Strings: A Course for Mathematicians*, Pierre Deligne, Pavel Etingof, Daniel S. Freed, Lisa C. Jeffrey, David Kazhdan, John W. Morgan, David R. Morrison, and Edward Witten, eds., vol. 1, Providence, RI: American Mathematical Society/Institute for Advanced Study, 1999, pp. 137–225.

DEWITT (Bryce S.)
[1957] Dynamical theory in curved spaces. I. A review of the classical and quantum action principles, *Reviews of Modern Physics*, 29 (1957), pp. 377–397.

[1964] Dynamical theory of groups and fields, in *Relativity, Groups and Topology* (Les Houches, 1963), C. DeWitt and B. DeWitt, eds., New York, London: Gordon and Breach, 1964, pp. 585–820; corrected and expanded edition, *Dynamical Theory of Groups and Fields*, New York, London: Gordon and Breach, 1965.

DICK (Auguste)
[1970] *Emmy Noether 1882–1935*, Kurze Mathematiker-Biographien, Beihefte zur Zeitschrift "Elemente der Mathematik," Suppléments à la "Revue de mathématiques élémentaires," no. 13, Basel: Birkhäuser, 1970.

[1981] *Emmy Noether 1882–1935*, English translation by Heidi I. Blocher, Boston, Basel, Stuttgart: Birkhäuser, 1981.

DICKEY (Leonid A.)
[1991] *Soliton Equations and Hamiltonian Systems*, Singapore: World Scientific, 1991; 2nd ed., revised, 2003.

[1994] Field-theoretical (multi-time) Lagrange–Hamilton formalism and integrable equations, in *Lectures on Integrable Systems*, O. Babelon, P. Cartier and Y. Kosmann-Schwarzbach, eds., Singapore: World Scientific, 1994, pp. 103–161.

DIEUDONNÉ (Jean), CARRELL (James B.)
[1971] *Invariant Theory, Old and New*, New York: Academic Press, 1971.

DIRAC (P. A. M.),
The Collected Works of P.A.M. Dirac, 1924–1948, R. H. Dalitz, ed., Cambridge, New York: Cambridge University Press, 1995.

[1936] Does conservation of energy hold in atomic processes? *Nature*, 137 (1936), pp. 288–289; *The Collected Works of P.A.M. Dirac*, pp. 803–804.

DONCEL (Manuel G.), HERMANN (Armin), MICHEL (Louis), PAIS (Abraham), eds.
[1987] *Symmetries in Physics (1600–1980)*, 1st International Meeting on the History of Scientific Ideas, Sant Feliu de Guíxols, Catalonia, Espagne (1983), Bellaterra (Barcelona): Seminari d'Història de les Ciències, Universitat Autònoma de Barcelona, 1987.

DORFMAN (Irene Ya.)
[1993] *Dirac Structures and Integrability of Nonlinear Evolution Equations*, Nonlinear Science: Theory and Applications, Chichester: John Wiley & Sons, 1993.

DROBOT (Stefan), RYBARSKI (Adam)
[1958] A variational principle of hydrodynamics, *Archive for Rational Mechanics and Analysis*, 2 (1958), pp. 393–410.

DUBREIL (Paul)
[1986] Emmy Noether, *Cahiers du séminaire d'histoire des mathématiques*, 7 (1986), pp. 15–27.

EARMAN (John), GLYMOUR (Clark)
[1978] Einstein and Hilbert: Two months in the history of general relativity, *Archive for History of Exact Sciences*, 19 (1978), pp. 291–308.

EARMAN (John), JANSSEN (Michel), NORTON (John D.), eds.
[1993] *The Attraction of Gravitation: New Studies in the History of General Relativity*, Einstein Studies, vol. 5, Boston: Birkhäuser, 1993.

EDELEN (Dominic G. B.),
[1969] *Nonlocal Variations and Local Invariance of Fields*, Modern Analytic and Computational Methods in Science and Mathematics, vol. 19, New York: American Elsevier Publishing Co., Inc., 1969.

EHRENFEST (Paul)
[1916] Over adiabatische veranderingen van een stelsel in verband met de theorie der quanta, *Koninklijke Akademie van Wetenschappen te Amsterdam, Verslagen van de Gewone Vergaderingen der Wis- en Natuurkundige Afdeeling*, 25 (1916–1917), pp. 412–433; English translation, On adiabatic changes of a system in connection with the quantum theory, *Koninklijke Akademie van Wetenschappen te Amsterdam, Proceedings of the Section of Sciences*, 19 (1916–1917), pp. 576–597.

EHRESMANN (Charles)
[1951] Les prolongements d'une variété différentiable, I: Calcul des jets, prolongement principal, *Comptes rendus de l'Académie des sciences de Paris*, 233 (1951), pp. 598–600.

EINSTEIN (Albert)
The Collected Papers of Albert Einstein, John Stachel *et al.*, eds., Princeton: Princeton University Press, 1987–.
Œuvres choisies, 6 vol., translated and edited by Françoise Balibar *et al.*, [Paris]: Seuil and Éditions du CNRS, 1989–1993.
[1907] Über das Relativitätsprinzip und die aus demselben gezogenen Folgerungen, *Jahrbuch der Radioaktivität und Elektronik*, 4 (1907), pp. 411–462; *Collected Papers* 2 (1989), no. 47, pp. 432–488; English translation, On the relativity principle and the conclusions drawn from it, *Collected Papers* 2 (English) (1997), pp. 252–311.
[1915] Die Feldgleichungen der Gravitation, *Sitzungsberichte der Königlich-preussischen Akademie der Wissenschaften zu Berlin* (1915), pp. 844–847 (meeting of 25 November 1915); *Collected Papers* 6 (1996), no. 25, pp. 244–249; English translation, The field equations of gravitation, *Collected Papers* 6 (English) (1997), pp. 117–120.
[1916a] Die Grundlage der allgemeinen Relativitätstheorie, *Annalen der Physik*, 49 (1916), pp. 769–822; reprint, Leipzig: J. A. Barth, 1916; *Collected Papers* 6 (1996), no. 30, pp. 283–339; English translation, The foundation of the general theory of relativity, in Lorentz *et al.* [1923], pp. 109–164, and in *Collected Papers* 6 (English) (1997), pp. 146–200; French translation, *Œuvres choisies* 2, pp. 179–227.
[1916b] Hamiltonsches Prinzip und allgemeine Relativitätstheorie, *Sitzungsberichte der Königlich-preussischen Akademie der Wissenschaften zu Berlin* (1916), pp. 1111–1116 (meeting of 26 October 1916); *Collected Papers* 6 (1996), no. 41, pp. 409–416; English translation, Hamilton's principle and the general theory of relativity, in Lorentz *et al.* [1923], pp. 165–173, and in *Collected Papers* 6 (English) (1997), pp. 240–246; French translation by Marc Lachièze-Rey in *Sur les épaules des géants*, Stephen Hawking, ed., Paris: Dunod, 2003, pp. 907–912.
[1918] Der Energiesatz in der allgemeinen Relativitätstheorie, *Sitzungsberichte der Königlich-preussischen Akademie der Wissenschaften zu Berlin* (1918), pp. 448–459 (meeting of 16 May 1918); *Collected Papers* 7 (2002), no. 9, pp. 63–77; English translation, The law of energy conservation in the general theory of relativity, *Collected Papers* 7 (English) (2002), pp. 47–61.

EINSTEIN (Albert), BESSO (Michele)
[1972] *Correspondance 1903–1955*, Translation into French, notes and introduction by Pierre Speziali, with notes on relativity by M.-A. Tonnelat, Paris: Hermann, 1972.

EISENSTAEDT (Jean), KOX (Anne J.), eds.
[1992] *Studies in the History of General Relativity*, vol. 3, Boston: Birkhäuser, 1992.

ELKANA (Yehuda)
[1974] *The Discovery of the Conservation of Energy*, London: Hutchinson Educational, 1974.

EL'SGOL'C (Lev E.)
[1952] *Variacionnoie Ischislenie* [Russian], Moscow, Leningrad: Gosudarstvennoe Izdateltsvo Tekhniko-teoreticheskoj Literatury, 1952; 2nd ed., 1958; English translation, L. E. Elsgolc, *Calculus of Variations*, Oxford: Pergamon Press, 1961.

ENCYCLOPÄDIE DER MATHEMATISCHEN WISSENSCHAFTEN
[1922] *Encyclopädie der mathematischen Wissenschaften*, Leipzig: B. G. Teubner, vol. 3, W. F. Meyer and H. Mohrmann, eds., 1922.

ENGEL (Friedrich)
[1916] Über die zehn allgemeinen Integrale der klassischen Mechanik, *Göttinger Nachrichten* (1916), Part 2, pp. 270–275.
[1917] Nochmals die allgemeinen Integrale der klassischen Mechanik, *Göttinger Nachrichten* (1917), Part 2, pp. 189–198.

ESHELBY (John D.)
[1975] The elastic energy-momentum tensor, *Journal of Elasticity*, 5 (1975), pp. 321–335.

EULER (Leonhard)
Opera Omnia, Leipzig, Berlin: B. G. Teubner, 1911-.
[1744] *Methodus inveniendi lineas curvas, maximi minimive proprietate gaudentes*, Lausanne, Geneva: Marc Michel Bousquet, 1744; *Opera Omnia*, Series I, vol. 24.
[1766] Elementa calculi variationum, *Novi Commentarii academiae scientiarum Petropolitanae*, 10, 1766, pp. 51–93; *Opera Omnia*, Series I, vol. 25, pp. 141–176.

FADDEEV (Ludwig D.)
[1982] The energy problem in Einstein's theory of gravitation [in Russian], *Uspekhi Fizicheskikh Nauk*, 136, no. 3 (1982), pp. 435–457; English translation, *Soviet Physics Uspekhi*, 25 (1982), no. 3, pp. 130–142.
[1987] 30 years in mathematical physics [in Russian], *Trudy Matematicheskovo Instituta imeni V. A. Steklova*, 176 (1987), pp. 4–29; English translation, *Proceedings of the Steklov Institute of Mathematics* (1988), no. 3, pp. 3–28; reprinted in *40 Years in Mathematical Physics*, World Scientific Series in 20th Century Mathematics, vol. 2, River Edge, NJ: World Scientific, 1995, pp. 1–28.

FEYNMAN (Richard P.), LEIGHTON (Robert B.), SANDS (Matthew)
[1963] *Lectures on Physics*, 3 vol., Reading, MA: Addison-Wesley, 1963–1965.

FISHER (Charles S.)
[1966] The death of a mathematical theory: a study in the sociology of knowledge, *Archive for History of Exact Sciences*, 3 (1966), pp. 137–159.

FLETCHER (John George)
[1960] Local conservation laws in generally covariant theories, *Reviews of Modern Physics*, 32 (1) (1960), pp. 65–87.

FOCK [Fok] (Vladimir Aleksandrovich)
[1955] *Teorija prostranstva, vremeni i tjagotenija* [Russian], Moscow: Gosudarstvennoe Izdateltsvo Tekhniko-teoreticheskoj Literatury, 1955; 2nd ed., revised, 1961; English translation by N. Kemmer, *The Theory of Space, Time and Gravitation*, London, New York, Paris, Los Angeles: Pergamon Press, 1959; 2nd ed., 1964.

FOGARTY (John)
[2001] On Noether's bound for polynomial invariants of a finite group, *Electronic Research Announcements*, American Mathematical Society, 7 (2001).

FOKAS (Athanassios S.), FUCHSSTEINER (Benno)
[1980] On the structure of symplectic operators and hereditary symmetries, *Lettere al Nuovo Cimento*, 2nd series, 28 (1980), pp. 299–303.

FOKKER (Adriaan Daniel)
[1917] De virtueele verplaatsingen van het elektromagnetische en van het zwaartekrachtsveld bij de toepassing van het variatiebeginsel van Hamilton, *Koninklijke Akademie van Wetenschappen te Amsterdam, Verslagen van de Gewone Vergaderingen der Wis- en Natuurkundige Afdeeling*, 25 (1916–1917), pp. 1067–1084; English translation, The virtual displacements of the electromagnetic and of the gravitational field in applications of Hamilton's variation principle, *Koninklijke Akademie van Wetenschappen te Amsterdam, Proceedings of the Section of Sciences*, 19 (1916–1917), pp. 968–984.

FORSYTH (Andrew Russel)
[1927] *Calculus of Variations*, London: Cambridge University Press, 1927.

FULP (Ronald O.), LADA (Thomas J.), STASHEFF (James D.)
[2003] Noether's variational theorem II and the BV formalism (Proceedings of the 22nd Winter School "Geometry and Physics," Srní, 2002), *Rendiconti del Circolo Matematico di Palermo*, Series II, Supplement no. 71 (2003), pp. 115–126.

FULTON (Thomas), ROHRLICH (Fritz), WITTEN (Louis)
[1962] Conformal invariance in physics, *Reviews of Modern Physics*, 34 (1962), pp. 442–457.

FUNK (Paul)
[1962] *Variationsrechnung und ihre Anwendung in Physik und Technik*, Die Grundlehren der mathematischen Wissenschaften, vol. 94, Berlin, Göttingen, Heidelberg: Springer, 1962; 2nd ed., 1970.

GARCÍA (Pedro L.)
[1968] Geometria simplectica en la teoria clásica de campos, *Collectanea Mathematica*, 19 (1968), pp. 73–134.
[1974] The Poincaré–Cartan invariant in the calculus of variations, in *Symposia Mathematica*, vol. 14, London, New York: Academic Press, 1974, pp. 219–246.

GEL'FAND (Israel M.), DIKIĬ (DICKEY) (Leonid A.)
[1975] Asymptotic behaviour of the resolvent of Sturm-Liouville equations [in Russian], *Uspekhi Matematicheskikh Nauk*, 30 (1975), pp. 67–100; English translation, *Russian Mathematical Surveys*, 30 (1975), p. 77–113; reprinted in Novikov *et al.* [1981], pp. 13–49.
[1976] A Lie algebra structure in the formal calculus of variations [in Russian], *Funksional'nyi Analiz i Ego Prilozhenyia*, 10 (1976), pp. 18–25; English translation, A Lie algebra structure in a formal variational calculation, *Functional Analysis and Its Applications*, 10 (1976), pp. 16–22.

GEL'FAND (Israel M.), DORFMAN (Irene Ya.)
[1979] Hamiltonian operators and algebraic structures connected with them [in Russian], *Funksional'nyi Analiz i Ego Prilozhenyia*, 13 (1979), pp. 13–30; English translation, *Functional Analysis and Its Applications*, 13 (1979), pp. 248–262.

GEL'FAND (Israel M.), FOMIN (Sergej V.)
[1961] *Variacionnoe ischislenie* [Russian], Moscow: Gosudarstvennoe Izdatelstsvo fiziko-matematicheskoj literatury, 1961; English translation by Richard A. Silverman, *Calculus of Variations*, Englewood Cliffs, NJ: Prentice-Hall, 1963; Dover Publications, 2000.

GEL'FAND (Israel M.), MANIN (Yuri I.), SHUBIN (Mikhail A.)
[1976] Poisson brackets and the kernel of variational derivatives in the formal calculus of variations [in Russian], *Funksional'nyi Analiz i Ego Prilozhenyia*, 10 (1976), pp. 30–34; English translation, *Functional Analysis and Its Applications*, 10 (1976), pp. 274–278.

GERSTENHABER (Murray)
[1964] On the deformation of rings and algebras, *Annals of Mathematics* (2), 79 (1964), pp. 59–103.

GOLDBERG (Joshua N.)
[1953] Strong conservation laws and equations of motion in covariant field theories, *Physical Review*, 89 (1953), pp. 263–272.
[1980] Invariant transformations, conservation laws, and energy-momentum, in *General Relativity and Gravitation*, vol. 1, A. Held, ed., New York: Plenum Press, 1980, pp. 469–489.

GOLDSCHMIDT (Hubert), STERNBERG (Shlomo)
[1973] The Hamilton–Cartan formalism in the calculus of variations, *Annales de l'Institut Fourier (Grenoble)*, 23 (1973), pp. 203–267.

GOLDSTEIN (Catherine), RITTER (Jim)
[2003] The varieties of unity. Sounding unified theories (1920–1930), in *Revisiting the Foundations of Relativistic Physics*, Abhay Ashtekar, Robert Cohen, Don Howard, Jürgen Renn, Sahotra Sarkar, and Abner Shimony, eds., Boston Studies in the Philosophy of Science, vol. 234, Dordrecht: Kluwer, 2003, pp. 93–149.

GOLDSTEIN (Herbert)
[1950] *Classical Mechanics*, Reading, MA: Addison-Wesley, 1950; reprints in the World Student Series Edition starting from 1962, 8th reprint, 1971; 2nd ed., Reading, MA: Addison-Wesley, 1980; French translation by A. and C. Moubacher and Théo Kahan, Paris: Presses Universitaires de France, 1964.

GOLDSTEIN (Herbert), POOLES (Charles), SAFKO (John)
[2002] *Classical Mechanics*, 3rd ed., Reading, MA: Addison-Wesley, 2002.

GOLDSTINE (Herman Heine),
[1980] *A History of the Calculus of Variations from the 17th through the 19th Century*, Studies in the History of Mathematics and Physical Sciences, vol. 5, New York, Heidelberg, Berlin: Springer, 1980.

GORDAN (Paul)
[1885] *Vorlesungen über Invariantentheorie*, lectures edited by Georg Kerschensteiner, vol. 1, *Determinanten*; vol. 2, *Binäre Formen*, Leipzig: Teubner, 1885–1887; reprint, in one volume, New York: Chelsea, 1987.

GOTAY (Mark J.), NESTER (James M.), HINDS (George)
[1978] Presymplectic manifolds and the Dirac–Bergmann theory of constraints, *Journal of Mathematical Physics*, 19 (1978), pp. 2388–2399.

GRAY (Jeremy J.), ed.
[1999] *The Symbolic Universe, Geometry and Physics 1890–1930* (Milton Keynes, 1996), Oxford, New York: Oxford University Press, 1999.

GROESBERG (Sanford W.)
[1968] *Advanced Mechanics*, New York: Wiley, 1968.

GUIL GUERRERO (Francisco), MARTÍNEZ ALONSO (Luis)
[1980a] Sobre un teorema de Noether, *Anales de Física de la Real Sociedad Española de Física y Química*, 76 (1980), pp. 189–191.
[1980b] Generalized variational derivatives in field theory, *Journal of Physics A*, 13 (1980), pp. 689–700.

GUTH (Eugene)
[1970] Contribution to the history of Einstein's geometry as a branch of physics, in *Relativity* (Proceedings, Cincinatti, 1969), Moshe Carmeli, Stuart I. Fickler, and Louis Witten, eds., New York, London: Plenum Press, 1970, pp. 161–207.

HAMEL (Georg)
[1904a] Die Lagrange–Eulerschen Gleichungen der Mechanik, *Zeitschrift für Mathematik und Physik*, 50 (1) (1904), pp. 1–57.
[1904b] Über die virtuellen Verschiebungen in der Mechanik, *Mathematische Annalen*, 59 (1904), pp. 416–434.

HAVAS (Peter)
[1973] The connection between conservation laws and invariance groups: folklore, fiction and fact, *Acta Physica Austriaca*, 38 (1973), pp. 145–167.
[1990] Energy-momentum tensors in special and general relativity (Developments in general relativity, astrophysics and quantum theory, Jerusalem and Haifa, 1989), *Annals of the Israel Physical Society*, 9 (1990), pp. 131–153.

HAWKINS (Thomas)
[1998] From general relativity to group representations, the background to Weyl's papers of 1925–26, in *Matériaux pour l'histoire des mathématiques au XXᵉ siècle*, Actes du colloque à la mémoire de Jean Dieudonné (Nice 1996), Séminaires et Congrès, vol. 3, [Paris]: Société mathématique de France, 1998, pp. 67–100.
[2000] *Emergence of the Theory of Lie Groups, An Essay in the History of Mathematics 1869–1926*, Sources and Studies in the History of Mathematics and Physical Sciences, New York: Springer-Verlag, 2000.

HELLER (Jack)
[1951] Covariant transformation laws for the field equations, *Physical Review*, 81 (1951), pp. 946–948.

HELMHOLTZ (Hermann von)
[1884] Principien der Statik monocyklischer Systeme, *Journal für die reine und angewandte Mathematik*, 97 (1884), pp. 111–140.
[1887] Über die physikalische Bedeutung des Princips der kleinsten Wirkung, *Journal für die reine und angewandte Mathematik*, 100 (1887), pp. 137–166.

HENNEAUX (Marc), TEITELBOIM (Claudio)
[1992] *Quantization of Gauge Systems*, Princeton: Princeton University Press, 1992.

HERGLOTZ (Gustav)
Gesammelte Schriften, Göttingen: Vandenhoeck & Ruprecht, 1979.
[1911] Über die Mechanik des deformierbaren Körpers vom Standpunkte der Relativitätstheorie, *Annalen der Physik* (Leipzig), 4th series, 36 (1911), pp. 493–533; *Gesammelte Schriften*, pp. 258–298.
[1916] Zur Einsteinschen Gravitationstheorie, *Sitzungsberichte der naturforschenden Gesellschaft Leipzig. Mathematisch-physikalische Klasse*, 68 (1916), pp. 199–203; *Gesammelte Schriften*, pp. 356–360.

HERMANN (Robert E.)
[1965] E. Cartan's geometric theory of partial differential equations, *Advances in Mathematics*, 1 (1965), pp. 265–317.
[1968] *Differential Geometry and the Calculus of Variations*, Mathematics in Science and Engineering, vol. 49, New York, London: Academic Press, 1968.
[1970a] *Vector Bundles in Mathematical Physics*, vol. 1, New York: Benjamin, 1970.
[1970b] *Lie Algebras and Quantum Mechanics*, New York: Benjamin, 1970.
[1970c] *Lectures in Mathematical Physics*, I and II, New York: Benjamin, 1970 and 1972.
[1973] *Geometry, Physics and Systems*, New York: Marcel Dekker, 1973.

HILBERT (David)
Gesammelte Abhandlungen, 3 vol., Berlin: J. Springer, 1932, 1933, 1935; reprint, New York: Chelsea, 1965.

[1915] Die Grundlagen der Physik I, *Göttinger Nachrichten* (1915), pp. 395–407 (meeting of 20 November 1915).

[1917] Die Grundlagen der Physik II, *Göttinger Nachrichten* (1917), pp. 53–76 (meeting of 23 December 1916).

[1924] Die Grundlagen der Physik, *Mathematische Annalen*, 92 (1924), pp. 1–32; *Gesammelte Abhandlungen*, vol. 3, pp. 258–289.

HILBERT (David), KLEIN (Felix)

[1985] *Der Briefwechsel David Hilbert–Felix Klein (1886–1918)*, Günther Frei, ed., Arbeiten aus der Niedersächsischen Staats- und Universitätsbibliothek Göttingen, vol. 19, Göttingen: Vandenhoeck & Ruprecht, 1985.

HILL (Edward Lee)

[1951] Hamilton's principle and the conservation theorems of mathematical physics, *Reviews of Modern Physics*, 23 (1951), pp. 253–260.

HOUTAPPEL (R. M. F.), VAN DAM (Hendrik), WIGNER (Eugene P.)

[1965] The conceptual basis and use of the geometric invariance principles, *Reviews of Modern Physics*, 37 (1965), pp. 595–632.

HOWARD (Don), STACHEL (John), eds.

[1989] *Einstein and the History of General Relativity*, Einstein Studies, vol. 1, Boston: Birkhäuser, 1989.

HOWE (Roger)

[1988] "The Classical Groups" and invariants of binary forms, in *The Mathematical Heritage of Hermann Weyl*, Proceedings of Symposia in Pure Mathematics, vol. 48, R. O. Wells, Jr., ed., Providence, RI: American Mathematical Society, 1988, pp. 145–146.

IBRAGIMOV (Nail H.)

[1969] Invariant variational problems and conservation laws (remarks on E. Noether's theorem) [in Russian with English abstract], *Teoreticheskaya i Matematicheskaya Fizika*, 1 (3) (1969), pp. 350–359; English translation, *Theoretical and Mathematical Physics*, 1 (3) (1969), pp. 267–274.

[1976] Lie–Bäcklund groups and conservation laws [in Russian], *Doklady Akad. Nauk SSSR*, 230 (1) (1976), pp. 26–29; English translation, *Soviet Mathematics Doklady*, 17 (5) (1976), pp. 1242–1246.

[1977] Group-theoretical nature of conservation theorems, *Letters in Mathematical Physics*, 1 (1977), pp. 423–428.

[1979] The Noether identity [in Russian], *Dinamika Sploshnoj Sredy*, 38 (1979), Novosibirsk, pp. 26–32.

[1983] *Gruppy preobrazovannij v matematicheskoj fizike* [Russian], Moscow: Nauka, 1983; English translation, *Transformation Groups Applied to Mathematical Physics*, Dordrecht, Boston, Lancaster: D. Reidel Publishing Company, 1985.

ITZYKSON (Claude), ZUBER (Jean-Bernard)

[1980] *Quantum Field Theory*, New York: McGraw-Hill, 1980.

IWANENKO (Dmitri D.), SOKOLOV (Arsenyi A.)

[1953] *Klassische Feldtheorie*, Berlin: Akademie-Verlag, 1953.

JACOBI (Carl Gustav Jacob)

[1866] *Vorlesungen über Dynamik* (1842–1843), lectures edited by Alfred Clebsch, 2nd ed., revised, Berlin: G. Reimer, 1866; reprint, New York: Chelsea, 1969; English translation by K. Balagangadharan, *Jacobi's lectures on dynamics*, Biswarup Banerjee, ed., New Delhi: Hindustan Book Agency, 2009.

JACOBSON (Nathan)

[1962] *Lie Algebras*, New York, London: Interscience Publ., John Wiley & Sons, 1962; reprint, New York: Dover, 1979.

JANSSEN (Michel)
  [1992] H. A. Lorentz's attempt to give a coordinate-free formulation of the general theory of relativity, in Eisenstaedt and Kox [1992], pp. 344–363.

JOHNSON (Harold H.)
  [1964a] Bracket and exponential for a new type of vector field, *Proceedings of the American Mathematical Society*, 15 (1964), pp. 432–437.
  [1964b] A new type of vector field and invariant differential systems, *Proceedings of the American Mathematical Society*, 15 (1964), pp. 675–678.

JULIA (Bernard), SILVA (Sebastian)
  [1998] Currents and superpotentials in classical gauge invariant theories, *Classical and Quantum Gravity*, 15 (1998), pp. 2173–2215.

KASTRUP (Hans A.)
  [1983] Canonical theories of Lagrangian dynamical systems in physics, *Physics Reports*, 101, nos. 1–2 (1983).
  [1987] The contributions of Emmy Noether, Felix Klein and Sophus Lie to the modern concept of symmetries in physical systems, in Doncel *et al.* [1987], pp. 113–158.

KERBRAT-LUNC (Hélène)
  [1964] Introduction mathématique à l'étude du champ de Yang–Mills sur un espace-temps courbe, *Comptes rendus de l'Académie des sciences de Paris*, 259 (1964), pp. 3449–3450.

KHUKHUNASHVILI [Huhunašvili] (Zaur V.)
  [1968] The Lagrangian formalism and the theory of symmetry. I [in Russian], *Izvestiya Vysshikh Uchebnykh Zavedenij. Fizika*, 1968, no. 11, pp. 7–16; English translation, *Soviet Physics Journal*, 11 (11) (1968), pp. 1–10.
  [1971] The symmetry of the differential equations of field theory [in Russian], *Izvestiya Vysshikh Uchebnykh Zavedenij. Fizika*, 1971, no. 3, pp. 95–103; English translation, The symmetry of the differential equations of field theory, *Soviet Physics Journal*, 14 (3) (1971), pp. 365–371.

KIJOWSKI (Jerzy), TULCZYJEW (Włodzimierz M.)
  [1979] *A Symplectic Framework for Field Theories*, Lecture Notes in Physics, 107, Berlin, Heidelberg, New York: Springer-Verlag, 1979.

KIMBERLING (Clark H.)
  [1972] Emmy Noether, *American Mathematical Monthly*, 79, no. 2 (1972), pp. 136–149. Addendum, *ibid.*, 79, no. 7 (1972), p. 755.
  [1981] Emmy Noether and her influence, in Brewer and Smith [1981], pp. 3–61.

KISHENASSAMY (S.)
  [1993] Variational derivation of Einstein's equations, in Earman, Janssen and Norton [1993], pp. 185–205.

KLEIN (Felix)
  Gesammelte mathematische Abhandlungen, 3 vol., Berlin: J. Springer, 1921–1923; reprint, Berlin, Heidelberg, New York: Springer, 1973.
  [1872] *Programm zum Eintritt in die philosophische Fakultät und den Senat der k. Friedrich-Alexanders-Univeristät zu Erlangen, Vergleichende Betrachtungen über neuere geometrische Forschungen* (das Erlanger Programm), Erlangen: A. Deichert, 1872; reprinted in *Mathematische Annalen*, 43 (1893), pp. 63–100, and in *Gesammelte mathematische Abhandlungen*, vol. 1, pp. 460–497; Italian translation by Gino Fano, Considerazioni comparative intorno a ricerche geometriche recenti, *Annali di Matematica Pura ed Applicata*, series 2, 17 (1889), pp. 307–343; French translation by Henri Eugène Padé, Considérations comparatives sur les recherches géométriques modernes, *Annales de l'École Normale* (3), 8 (1891), pp. 87–102 and 173–199; reprint, with a preface by Jean Dieudonné, Paris: Gauthier-Villars, 1974; reprint, Paris: Éditions Jacques Gabay, 1991;

English translation by M. W. Haskell, "Felix Klein, A comparative review of recent researches in geometry," *Bulletin of the New York Mathematical Society*, 2 (1892–1893), pp. 215–249.

[1910] Über die geometrischen Grundlagen der Lorentzgruppe, *Jahresbericht der Deutschen Mathematiker-Vereinigung*, 19 (1910), pp. 281–300; reprint, *Physikalische Zeitschrift*, 12 (1911), pp. 17–27; *Gesammelte mathematische Abhandlungen*, vol. 1, pp. 533–552.

[1918a] Zu Hilberts erster Note über die Grundlagen der Physik, *Göttinger Nachrichten* (1917), pp. 469–482 (meeting of 25 January 1918); *Gesammelte mathematische Abhandlungen*, vol. 1, pp. 553–567.

[1918b] Über die Differentialgesetze für die Erhaltung von Impuls und Energie in der Einsteinschen Gravitationstheorie, *Göttinger Nachrichten* (1918), pp. 171–189 (meeting of 19 July 1918); *Gesammelte mathematische Abhandlungen*, vol. 1, pp. 568–585.

[1918c] Über die Integralform der Erhaltungssätze und die Theorie der räumlich-geschlossenen Welt, *Göttinger Nachrichten* (1918), pp. 394–423 (meeting of 6 December 1918, printed end of January 1919); *Gesammelte mathematische Abhandlungen*, vol. 1, pp. 586–612.

[1927] *Vorlesungen über die Entwicklung der Mathematik im 19. Jahrhundert*, I and II, Grundlehren der mathematischen Wissenschaften, 24 and 25, Berlin: J. Springer, 1926 and 1927; English translation of volume I by M. Ackerman, in R. Hermann, *Interdisciplinary Mathematics*, 9, Brookline, MA: Math. Sci. Press, 1979.

KNESER (Adolf)

[1900] *Lehrbuch der Variationsrechnung*, Braunschweig: Vieweg, 1900; 2nd edition, 1925.

[1917] Transformationsgruppen und Variationsrechnung, *Journal für die reine und angewandte Mathematik*, 147 (1917), pp. 54–66.

[1918] Kleinste Wirkung und Galileische Relativität, *Mathematische Zeitschrift*, 2 (1918), pp. 326–349.

KOLÁŘ (Ivan)

[1984] A geometrical version of the higher order Hamilton formalism in fibred manifolds, *Journal of Geometry and Physics*, 1 (1984), pp. 127–137.

KOLÁŘ (Ivan), MICHOR (Peter W.), SLOVÁK (Jan)

[1993] *Natural Operations in Differential Geometry*, Berlin: Springer-Verlag, 1993.

KOMAR (Arthur)

[1959] Covariant conservation laws in general relativity, *Physical Review* (2), 113 (1959), pp. 934–936.

[1967] Gravitational superenergy as a generator of canonical transformation, *Physical Review*, 164 (1967), pp. 1595–1599.

KOMOROWSKI (Jan)

[1968] A modern version of the E. Noether's theorems in the calculus of variations, I and II, *Studia Mathematica*, 29 (1968), pp. 261–273 and 32 (1969), pp. 181–190.

KOSMANN-SCHWARZBACH (Yvette)

[1979] Generalized symmetries of nonlinear partial differential equations, *Letters in Mathematical Physics*, 3 (1979), pp. 395–404.

[1980] Vector fields and generalized vector fields on fibered manifolds, in *Geometry and Differential Geometry* (Proceedings of a Conference held at the University of Haifa, Israel, 1979), R. Artzy and I. Vaisman, eds., Lecture Notes in Mathematics, 792, Berlin, Heidelberg, New York: Springer-Verlag, 1980, pp. 307–355.

[1985] On the momentum mapping in field theory, in *Differential Geometric Methods in Mathematical Physics* (Proceedings, Clausthal 1983), H. D. Doebner and J. D. Hennig, eds., Lecture Notes in Mathematics, 1139, Berlin, Heidelberg, New York: Springer-Verlag, 1985, pp. 25–73.

[1987] Sur les théorèmes de Noether, in *Géométrie et physique* (Journées relativistes de Marseille-Luminy, 1985), Y. Choquet-Bruhat, B. Coll, R. Kerner, and A. Lichnerowicz, eds., Travaux en Cours, vol. 21, Paris: Hermann, 1987, pp. 147–160.

KOSTANT (Bertram)

[1970] Quantization and unitary representations, in *Lectures in Modern Analysis and Applications* III, C. T. Taam, ed., Lecture Notes in Mathematics, 170, Berlin, Heidelberg, New York: Springer-Verlag, 1970, pp. 87–208.

KOX (Anne J.)

[1992] General Relativity in the Netherlands (1915–1920), in Eisenstaedt and Kox [1992], pp. 39–56.

KRASIL'SHCHIK (Iosif S.), VINOGRADOV (Aleksandr M.), eds.

[1997] *Simmetrii i zakony sokhranenija uravnenij matematicheskoj fiziki* [Russian], Moscow: Factorial, 1997; English translation by A. M. Verbovetsky and I. S. Krasil'shchik, *Symmetries and Conservation Laws for Differential Equations of Mathematical Physics*, Providence, RI: American Mathematical Society, 1999.

KREYSZIG (Erwin)

[1994] On the calculus of variations and its major influences on the mathematics of the first half of our century, *American Mathematical Monthly*, 101, no. 7 (1994), pp. 674–678); reprinted in *The Genius of Euler: Reflections*, William Dunham, ed., *s.l.*: Mathematical Association of America, 2007, pp. 209–214.

KRUPKA (Demeter)

[1971] Lagrange theory in fibered manifolds, *Reports on Mathematical Physics*, 2 (1971), pp. 121–133.

[1973] Some geometric aspects of variational problems in fibered manifolds, *Folia Facultatis Scientiarum Naturalium Universitatis Purkynianae Brunensis*, 14 (10) (1973), 65 pp.

[1975] A geometric theory of ordinary first order variational problems in fibered manifolds, I. Critical sections, *Journal of Mathematical Analysis and Applications*, 49 (1975), pp. 180–206; II. Invariance, *ibid.*, pp. 469–476.

KRUPKA (Demeter), TRAUTMAN (Andrzej)

[1974] General invariance of Lagrangian structures, *Bulletin de l'Académie Polonaise des Sciences, Série Mathématique, Astronomie, Physique*, 22 (1974), pp. 207–211.

KRUPKOVA (Olga)

[2009] Noether Theorem, 90 years on, in *Geometry and Physics: XVII International Fall Workshop*, Fernando Etayo, Mario Fioravanti and Rafael Santamaria, eds., AIP Conference Proceedings 1130, American Institute of Physics, 2009.

KUPERSHMIDT (Boris A.)

[1980] Geometry of jet bundles and the structure of Lagrangian and Hamiltonian formalisms, in *Geometric Methods in Mathematical Physics* (Lowell, MA, 1979), Lecture Notes in Mathematics, 775, G. Kaiser and J. E. Marsden, eds., Berlin, Heidelberg, New York: Springer-Verlag, 1980.

LAGRANGE (Joseph Louis)

Œuvres de Lagrange, Paris: Gauthier-Villars, 1867–1892.

[1760] Essai d'une nouvelle méthode pour déterminer les maxima et les minima des formules intégrales indéfinies, *Miscellanea Taurinensia*, 2 (1760-1761); *Œuvres de Lagrange*, vol. 1, pp. 335–362.

[1777] Remarques générales sur le mouvement de plusieurs corps qui s'attirent mutuellement en raison inverse du carré des distances, *Nouveaux mémoires de l'Académie Royale des Sciences et Belles-Lettres de Berlin* (1777), pp. 155–172; *Œuvres de Lagrange*, vol. 4, pp. 401–418.

[1788] *Méchanique Analitique*, Paris: chez la Veuve Desaint, 1788.

[1811] *Mécanique Analytique*, 2nd ed., Paris: Mme Veuve Courcier, vol. 1, 1811; vol. 2, 1815; new ed., 2 vol., Paris: Albert Blanchard, 1965, based on the 3rd (1853–1855) and 4th (1888) eds.

LANCZOS (Cornelius)
[1949] *The Variational Principles of Mechanics*, Toronto: University of Toronto Press, 1949; reprint, 1952, 1957, 1960; 2nd ed., 1962, reprint, 1964; 3rd ed., 1966; 4th ed., 1970; reprint of the 4th ed., New York: Dover, 1986.
[1973] Emmy Noether and the calculus of variations, *Bulletin of the Institute of Mathematics and Its Applications* (Southend-on-Sea, Great Britain) (August 1973), pp. 253–258.

LANDAU (Lev D.), LIFSHITZ (Evgenij M.)
[1948] *Teoreticheskaja fizika*, vol. 2, *Teorija polja*, Moscow: Gosudarstvennoe Izdateltsvo Tekhniko-teoreticheskoj Literatury, 1948; 3rd ed., 1960; 6th ed., revised, 1973; English translation by Morton Hamermesh, *The Classical Theory of Fields*, Cambridge, MA: Addison–Wesley, 1951 (first edition, 1941); 2nd ed., revised, Oxford, London, Paris: Pergamon Press, 1962; 3rd ed., revised, translated from the 3rd ed. of *Teorija polja*, Oxford, New York: Pergamon Press, 1971; 4th ed., revised, translated from the 6th ed. of *Teorija polja*, Oxford [etc.]: Butterworth–Heinemann, 1975.

LEPAGE (Théodore H. J.)
[1936] Sur les champs géodésiques du calcul des variations, I and II, *Académie royale de Belgique, Bulletins de la classe des sciences*, 5th series, 22 (1936), pp. 716–729 and 1034–1046.

LEVI-CIVITA (Tullio)
Opere matematiche, 4 vol., Memorie e Note, Accademia Nazionale dei Lincei, Bologna: Nicola Zanichelli, 1954–1960.
[1896] Sulle transformazioni delle equazioni dinamiche, *Annali di Matematica*, 2nd series, 24 (1896), pp. 255–300; *Opere matematiche*, vol. 1, pp. 207–252.
[1925] *Lezioni di calcolo differenziale assoluto*, collected and edited by Enrico Persico, Rome: A. Stock, 1925; English translation by Marjorie Long, *The Absolute Differential Calculus (Calculus of Tensors)*, London and Glasgow: Blackie & Son, 1926.

LEVI-CIVITA (Tullio), AMALDI (Ugo)
[1923] *Lezioni di meccanica razionale*, 2 vol., Bologna: Nicola Zanichelli, 1923 and 1927.

LIBERMANN (Paulette), MARLE (Charles-Michel)
[1987] *Symplectic Geometry and Analytical Mechanics*, translated from French by Bertram E. Schwarzbach, Dordrecht: D. Reidel Publishing Company, 1987.

LICHNEROWICZ (André)
[1955] *Théories relativistes de la gravitation et de l'électromagnétisme. Relativité générale et théories unitaires*, Paris: Éditions Masson, 1955; reprint, Paris: Jacques Gabay, 2011.
[1957] *Théorie globale des connexions et des groupes d'holonomie*, Roma: Edizioni Cremonese, 1957; reprint, Paris: Dunod, 1962; Russian translation by Sergei P. Finikov, Moskow: Izdat. Inostr. Lit., 1960.

LIE (Sophus)
Gesammelte Abhandlungen, vol. 6, Leipzig: B. G. Teubner; Oslo: H. Aschehoug, 1927.
[1897a] Die Theorie der Integralinvarianten ist ein Korollar der Theorie der Differentialinvarianten, *Sitzungsberichte der Sächsischen Akademie der Wissenschaften zu Leipzig. Mathematisch-Naturwissenschaftliche Klasse* (1897), Part III, pp. 342–357; *Gesammelte Abhandlungen*, vol. 6, pp. 649–663.
[1897b] Über Integralinvarianten und ihre Verwertung für die Theorie der Differentialgleichungen, *Sitzungsberichte der Sächsischen Akademie der Wissenschaften zu Leipzig. Mathematisch-Naturwissenschaftliche Klasse* (1897), Part IV, pp. 369–410; *Gesammelte Abhandlungen*, vol. 6, pp. 664–701.

LIE (Sophus), ENGEL (Friedrich)
[1893] *Theorie der Transformationsgruppen*, Leipzig: B. G. Teubner, 3 vol., 1888, 1890, 1893; reprint, New York: Chelsea, 1970.

LIPKIN (Daniel M.)

[1964] Existence of a new conservation law in electromagnetic theory, *Journal of Mathematical Physics*, 5 (1964), pp. 696–700.

LOGAN (John David)

[1973] First integrals in the discrete variational calculus, *Aequationes Mathematicae*, 9 (1973), pp. 210–220.

[1974] On variational problems which admit an infinite continuous group, *Yokohama Mathematical Journal*, 22 (1974), pp. 31–42.

[1977] *Invariant Variational Principles*, Mathematics in Science and Engineering, vol. 138, New York: Academic Press, 1977.

LORENTZ (Hendrik A.)

Collected Papers, 9 vol., The Hague: Nijhoff, 1934–1939.

[1915] Het beginsel van Hamilton in Einstein's theorie der zwaartekracht, *Koninklijke Akademie van Wetenschappen te Amsterdam, Verslagen van de Gewone Vergaderingen der Wis- en Natuurkundige Afdeeling*, 23 (1915–1916), pp. 1013–1073; English translation, On Hamilton's principle of Einstein's theory of gravitation, *Koninklijke Akademie van Wetenschappen te Amsterdam, Proceedings of the Section of Sciences*, 19 (1916–1917), pp. 751–765; English translation reprinted in *Collected Papers*, vol. 5 (1937), pp. 229–245.

[1916] Over Einstein's theorie der zwaartekracht, I, II, and III, IV, *Koninklijke Akademie van Wetenschappen te Amsterdam, Verslagen van de Gewone Vergaderingen der Wis- en Natuurkundige Afdeeling*, 24 (1915–1916), pp. 1389–1402 and 1759–1774, and 25 (1916–1917), pp. 468–486 and 1380–1396; English translation, On Einstein's theory of gravitation, I, II, and III, IV, *Koninklijke Akademie van Wetenschappen te Amsterdam, Proceedings of the Section of Sciences*, 19 (1916–1917), pp. 1341–1354 and 1354–1369, and 20 (1917–1918), pp. 2–19 and 20–34; English translation reprinted in *Collected Papers*, vol. 5 (1937), pp. 246–313.

LORENTZ (Hendrik A.), EINSTEIN (Albert), MINKOWSKI (Hermann), WEYL (Hermann)

[1922] *Das Relativitätsprinzip*, eine Sammlung von Abhandlungen, mit einem Beitrag von H. Weyl und Anmerkungen von A. Sommerfeld, Vorwort von O. Blumenthal, 4th ed., Leipzig: Teubner, 1922; 5th ed., 1923; 6th ed., Stuttgart: Teubner, 1958. The earlier editions of *Das Relativitätsprinzip* (1913, 1915, and 1920) do not contain Weyl's contribution.

[1923] *The Principle of Relativity, a collection of original memoirs on the special and general theory of relativity, by H. A. Lorentz, A. Einstein, H. Minkowski and H. Weyl*, Arnold Sommerfeld, ed., English translation of the 4th German edition [1922] by W. Perrett and G. B. Jeffery, London: Methuen, 1923; reprint, New York: Dover, 1952.

LOVELOCK (David), RUND (Hanno)

[1975] *Tensors, Differential Forms, and Variational Principles*, Pure and Applied Mathematics, New York, London, Sydney: John Wiley & Sons, 1975; reprint, New York: Dover, 1989.

LUSANNA (Luca)

[1991] The second Noether theorem as the basis of the theory of singular Lagrangians and Hamiltonian constraints, *Rivista del Nuovo Cimento* (3), 14 (1991), no. 3.

MAC LANE (Saunders)

[1981] Mathematics at the University of Göttingen 1931–1933, in Brewer and Smith [1981], pp. 65–78.

MAGRI (Franco)

[1978] Sul legame tra simmetrie e leggi di conservazione nelle teorie di campo classiche, 4th National Congress of Theoretical and Applied Mechanics, AIMETA, Florence, 25–28 October 1978.

MANIN (Yuri I.)

[1978] Algebraic aspects of nonlinear differential equations [in Russian], *Itogi Nauki i Tekhniki, Sovremennye Problemy Matematiki*, 11 (1978), pp. 5–152; English translation, *Journal of Soviet Mathematics*, 11 (1979), pp. 1–122.

MANSFIELD (Elizabeth L.)

[2006] Noether's theorem for smooth, difference and finite element systems, in *Foundations of Computational Mathematics* (Santander, 2005), Luis M. Pardo, Allan Pinkus, Endre Süli, and Michael J. Todd, eds., London Mathematical Society Lecture Note Series, vol. 331, Cambridge: Cambridge University Press, 2006, pp. 230–254.

MANSFIELD (Elizabeth L.), HYDON (Peter E.)

[2001] On a variational complex for difference equations, in *The Geometrical Study of Differential Equations* (Washington, DC, 2000), Joshua A. Leslie and Thierry P. Robart, eds., Contemporary Mathematics, vol. 285, Providence, RI: American Mathematical Society, 2001, pp. 121–129.

MARKOW (Moisei A.)

[1936] Zur Diracschen Theorie des Elektrons, *Physikalische Zeitschrift der Sowjetunion*, 10 (1936), pp. 773–808.

MARLE (Charles-Michel)

[1983] Lie group actions on a canonical manifold, in *Symplectic Geometry*, A. Crumeyrolle and J. Grifone, eds., Research Notes in Mathematics, vol. 8, Boston: Pitman, 1983, pp. 144–166.

MARSDEN (Jerry [Jerrold E.])

[1974] *Applications of Global Analysis in Mathematical Physics*, Boston, MA: Publish or Perish, inc., 1974.

MARSDEN (Jerrold E.), RATIU (Tudor S.)

[1999] *Introduction to Mechanics and Symmetry*, 2nd ed., Texts in Applied Mathematics, vol. 17, New York, Berlin, Heidelberg: Springer, 1999.

MARTÍNEZ ALONSO (Luis)

[1979] On the Noether map, *Letters in Mathematical Physics*, 3 (1979), pp. 419–424.

MASCHKE (Heinrich)

[1900] A new method of determining the differential parameters and invariants of quadratic differential quantics, *Transactions of the American Mathematical Society*, 1 (1900), pp. 197–204.

[1903] A symbolic treatment of the theory of invariants of differential quantics of n variables, *Transactions of the American Mathematical Society*, 4 (1903), pp. 445–469.

MEHRA (Jagdish)

[1974] *Einstein, Hilbert and the Theory of Gravitation: Historical Origins of General Relativity Theory*, Dordrecht: D. Reidel Publishing Company, 1974; first published (without index) in *The Physicist's Conception of Nature*, J. Mehra, ed., Dordrecht: D. Reidel Publishing Company, 1973, pp. 92–178.

MELVIN (M. Avramy)

[1960] Elementary particles and symmetry principles, *Reviews of Modern Physics*, 32 (1960), pp. 477–518.

MØLLER (Christian)

[1958] On the localization of the energy of a physical system in the general theory of relativity, *Annals of Physics*, 4 (1958), pp. 347–371.

NE'EMAN (Yuval)

[1999] The impact of Emmy Noether's theorems on XXIst century physics, in Teicher [1999], pp. 83–101.

NOETHER (Emmy)

Gesammelte Abhandlungen / Collected papers, N. Jacobson, ed., Berlin, Heidelberg, New York: Springer-Verlag, 1983.

[1907] Über die Bildung des Formensystems der ternären biquadratischen Form, *Sitzungsberichte der Physikalisch-medizinischen Societät zu Erlangen*, 39 (1907), pp. 176–179; *Abhandlungen*, pp. 27–30.

[1908] Über die Bildung des Formensystems der ternären biquadratischen Form, *Journal für die reine und angewandte Mathematik*, 134 (1908), pp. 23–90, with two tables; *Abhandlungen*, pp. 31–99.

[1910] Zur Invariantentheorie der Formen von *n* Variabeln, *Jahresbericht der Deutschen Mathematiker-Vereinigung*, 19 (1910), pp. 101–104; *Abhandlungen*, pp. 100–103.

[1911] Zur Invariantentheorie der Formen von *n* Variabeln, *Journal für die reine und angewandte Mathematik*, 139 (1911), pp. 118–154; *Abhandlungen*, pp. 104–140.

[1913] Rationale Funktionenkörper, *Jahresbericht der Deutschen Mathematiker-Vereinigung*, 22 (1913), pp. 316–319; *Abhandlungen*, pp. 141–144.

[1915] Körper und Systeme rationaler Funktionen, *Mathematische Annalen*, 76 (1915), pp. 161–196; *Abhandlungen*, pp. 145–180.

[1916a] Der Endlichkeitssatz der Invarianten endlicher Gruppen, *Mathematische Annalen*, 77 (1916), pp. 89–92; *Abhandlungen*, pp. 181–184.

[1916b] Über ganze rationale Darstellung der Invarianten eines Systems von beliebig vielen Grundformen, *Mathematische Annalen*, 77 (1916), pp. 93–102; *Abhandlungen*, pp. 185–194.

[1916c] Die allgemeinsten Bereiche aus ganzen transzendenten Zahlen, *Mathematische Annalen*, 77 (1916), pp. 103–128; *Abhandlungen*, pp. 195–220.

[1916d] Die Funktionalgleichungen der isomorphen Abbildung, *Mathematische Annalen*, 77 (1916), pp. 536–545; *Abhandlungen*, pp. 221–230.

[1918a] Gleichungen mit vorgeschriebener Gruppe, *Mathematische Annalen*, 78 (1918), pp. 221–229; *Abhandlungen*, pp. 231–239.

[1918b] Invarianten beliebiger Differentialausdrücke, *Göttinger Nachrichten* (1918), pp. 37–44 (presented by F. Klein at the meeting of 25 January 1918); *Abhandlungen*, pp. 240–247. Abstract by the author in *Jahrbuch über die Fortschritte der Mathematik*, 46 (1916–1918), vol. 1, IV.9 (*Gewöhnliche Differentialgleichungen*), p. 675.

[1918c] Invariante Variationsprobleme, *Göttinger Nachrichten* (1918), pp. 235–257 (presented by F. Klein at the meeting of 26 July 1918); *Abhandlungen*, pp. 248–270. Abstract by the author in *Jahrbuch über die Fortschritte der Mathematik*, 46 (1916–1918), vol. 1, IV.15 (*Variationsrechnung*), p. 770.

[1919] Die Endlichkeit des Systems der ganzzahligen Invarianten binärer Formen, *Göttinger Nachrichten* (1919), pp. 138–156 (presented by F. Klein at the meeting of 27 March 1919); *Abhandlungen*, pp. 293–311.

[1922] Formale Variationsrechnung und Differentialinvarianten, in Weitzenböck [1922], pp. 68–71; *Abhandlungen*, pp. 405–408 (with the reference to II.3 in *Encyklopädie der mathematischen Wissenschaften* given by mistake instead of III.3) [This article is attributed to Noether].

[1923] Algebraische und Differentialinvarianten, *Jahresbericht der Deutschen Mathematiker-Vereinigung*, 32 (1923), pp. 177–184; *Abhandlungen*, pp. 436–443 (the faulty title is "Algebraische und Differentialvarianten").

[1929] Hyperkomplexe Grössen und Darstellungstheorie, *Mathematische Zeitschrift*, 30 (1929), pp. 641–692; *Abhandlungen*, pp. 563–614.

[1971] Invariant variation problems, English translation of [1918c] by M. A. Tavel, *Transport Theory and Statistical Physics*, 1 (3) (1971), pp. 186–207.

NOETHER (Herman D.)
[1993] Fritz Alexander Noether—Opfer zweier Diktaturen, *Physikalische Blätter*, 49 (1993), no. 9, p. 815.

NORDSTRÖM (Gunnar)
[1917] De gravitatsietheorie van Einstein en de mechanica der continua van Herglotz, *Koninklijke Akademie van Wetenschappen te Amsterdam, Verslagen van de Gewone Vergaderingen der Wis- en Natuurkundige Afdeeling*, 25 (1916–1917), pp. 836–843; English translation, Einstein's theory of gravitation and Herglotz's mechanics of continua, *Koninklijke Akademie van Wetenschappen te Amsterdam, Proceedings of the Section of Sciences*, 19 (1916–1917), pp. 884–891.

NORTON (John D.)

[1984] How Einstein found his field equations, 1912–1915, *Historical Studies in the Physical Sciences*, 14 (1984), pp. 253–316; reprinted in Howard and Stachel [1989], pp. 101–159.

[1999] Geometries in collision: Einstein, Klein and Riemann, in Gray [1999], pp. 128–144.

NOVIKOV (Sergei P.) *et al.*

[1981] *Integrable Systems*, London Mathematical Society Lecture Notes Series, vol. 60, Cambridge: Cambridge University Press, 1981.

OLVER (Peter J.)

[1977] Evolution equations possessing infinitely many symmetries, *Journal of Mathematical Physics*, 18 (1977), pp. 1212–1215.

[1984] Conservation laws in elasticity, I. General results, *Archive for Rational Mechanics and Analysis*, 85 (1984), pp. 111–129.

[1986a] *Applications of Lie Groups to Differential Equations*, Graduate Texts in Mathematics, vol. 107, New York: Springer-Verlag, 1986; 2nd ed., revised, 1993.

[1986b] Noether's theorems and systems of Cauchy-Kovalevskaya type, in *Nonlinear Systems of Partial Differential Equations in Applied Mathematics* (Proceedings of the SIAM-AMS Summer Seminar, Santa Fe, NM), Basil Nicolaenko, Darryl D. Holm, and James M. Hyman, eds., Lectures in Applied Mathematics, vol. 23, Part 2, Providence, RI: American Mathematical Society, 1986, pp. 81–104.

[1995] *Equivalence, Invariants, and Symmetry*, Cambridge: Cambridge University Press, 1995.

[1999] *Classical Invariant Theory*, London Mathematical Society student texts, vol. 44, Cambridge: Cambridge University Press, 1999.

O'RAIFEARTAIGH (Lochlainn)

[1997] *The Dawning of Gauge Theory*, Princeton: Princeton University Press, 1997.

OVSJANNIKOV (Lev Vasil'evich)

[1978] *Gruppovoj analiz differentsial'nykh uravnenij* [Russian], Moscow: Nauka (Glavnaja Red. Fiziko-matematicheskoj Literatury), 1978; English translation by William F. Ames, *Group Analysis of Differential Equations*, New York: Academic Press, 1982.

PAIS (Abraham)

[1982] *"Subtle is the Lord ... " The Science and the Life of Albert Einstein*, Oxford: Clarendon Press, 1982.

[1987] Conservation of energy, in Doncel *et al.* [1987], pp. 361–377.

PARS (Leopold Alexander)

[1962] *An Introduction to the Calculus of Variations*, London: Heinemann,1962.

PARSHALL (Karen Hunger)

[1989] Toward a history of nineteenth-century invariant theory, in *The History of Modern Mathematics*, vol. 1, David E. Rowe and John McCleary, eds., Boston: Academic Press, 1989, pp. 157–206.

PAULI (Wolfgang)

Collected Scientific Papers, 2 vol., R. Kronig and Victor F. Weisskopf, eds., New York, London, Sydney: Interscience Publ., a division of J. Wiley, 1964.

[1921] Relativitätstheorie, *Encyklopädie der mathematischen Wissenschaften*, vol. V.2, par. 19, Leipzig: B. G. Teubner, 1921, pp. 539–775; reprinted as a monograph, with a preface by Arnold Sommerfeld, Leipzig: Teubner, 1921; *Collected Scientific Papers*, vol. 1, pp. 1–237.

[1941] Relativistic field theories of elementary particles, *Reviews of Modern Physics*, 13 (1941), pp. 203–232; *Collected Scientific Papers*, vol. 2, pp. 923–952.

[1958] *Theory of Relativity*, English translation of [1921] by G. Field, with additional notes by the author, London, New York, Paris, Los Angeles: Pergamon Press, 1958; reprint, New York: Dover, 1981; additional notes reprinted in *Collected Scientific Papers*, vol. 1, pp. 238–263.

[1979] *Wissenschaftlicher Briefwechsel mit Bohr, Einstein, Heisenberg u.a.*, Band I: 1919–1929. *Scientific Correspondence with Bohr, Einstein, Heisenberg, a.o.*, vol. 1, 1919–1929, Armin Hermann, Karl von Meyenn, and Victor F. Weisskopf, eds., New York, Heidelberg, Berlin: Springer-Verlag, 1979.

POINCARÉ (Henri)
[1899] *Les méthodes nouvelles de la mécanique céleste*, vol. 3, Paris: Gauthier-Villars, 1899.

POLAK (Lev S.)
[1959] *Variacionnye principy mekhaniki* [Russian] (Variational Principles of Mechanics), Moscow: Gosudarstvennoe Izdatelstsvo Fiziko-matematicheskoj Literatury, 1959.

POLAK (Lev S.), VIZGIN (Vladimir P.)
[1979] The Noether theorem in the history of physics [in Russian], *Istorija i metodologija estestvennykh nauk* (*History and methodology of the natural sciences*), 22: *Fizika* (*Physics*), Moscow: Moskov. Gos. Univ., 1979, pp. 99–110.

PROCESI (Claudio)
[1999] 150 years of invariant theory, in Teicher [1999], pp. 6–21.

REID (Constance)
[1970] *Hilbert*, Berlin, Heidelberg, New York: Springer-Verlag, 1970.

RICCI-CURBASTRO (Gregorio)
Opere, 2 vol., Roma: Edizioni Cremonese, 1956–1957.
[1892] Résumé de quelques travaux sur les systèmes variables de fonctions associés à une forme différentielle quadratique, *Bulletin des Sciences Mathématiques*, 2nd series, 16 (1892), pp. 167–189; reprinted in *Opere*, vol. 1, pp. 288–310.

RICCI-CURBASTRO (Gregorio), LEVI-CIVITA (Tullio)
[1900] Méthodes de calcul différentiel absolu et leurs applications, *Mathematische Annalen*, 54 (1900), pp. 125–201; reprinted in Levi-Civita, *Opere matematiche*, vol. 1, pp. 479–559.

ROMAN (Paul)
[1955] Erhaltungsgesetze und quantenmechanische Operatoren, *Acta Physica Academiae Scientiarum Hungaricae*, 5 (1955), pp. 143–158.
[1960] *Theory of Elementary Particles*, Amsterdam: North-Holland, 1960; 2nd ed., 1964.
[1969] *Introduction to Quantum Field Theory*, New York: Wiley, 1969.

ROQUETTE (Peter)
[2008] Emmy Noether and Hermann Weyl, in *Groups and Analysis: The Legacy of Hermann Weyl*, Katrin Tent, ed., London Mathematical Society Lecture Note Series, vol. 354, Cambridge: Cambridge University Press, pp. 285–326,

ROSEN (Joe)
[1972] Noether's theorem in classical field theory, *Annals of Physics*, 69 (1972), pp. 349–363.
[1981] Resource letter SP-2: Symmetry and group theory in physics, *American Journal of Physics*, 49 (1981), pp. 304–319.

ROSENFELD (Léon)
[1979] *Selected Papers of Léon Rosenfeld*, Robert S. Cohen and John J. Stachel, eds., with an introduction by Stefan Rozental, Boston Studies in the Philosophy of Science, vol. 21, Dordrecht, Boston, London: Reidel, 1979.
[1930] Zur Quantelung der Wellenfelder, *Annalen der Physik* (Leipzig), 5 (1930), pp. 113–152.
[1940] Sur le tenseur d'impulsion-énergie, *Académie Royale de Belgique, Classe des Sciences, Mémoires*, 18 (1940), fasc. 6, 30 pp.; English translation in [1979], pp. 711–735.

ROUTH (Edward John)
[1877] *Stability of a given state of motion*, London: Macmillan and Co., 1877; reprinted in *Stability of Motion*, A. T. Fuller, ed., London: Taylor and Francis, 1975.

ROWE (David E.)

[1999] The Göttingen response to general relativity and Emmy Noether's theorems, in Gray [1999], pp. 189–233.

[2000] Episodes in the Berlin–Göttingen rivalry, 1870–1930, *Mathematical Intelligencer*, 22(1) (2000), pp. 60–69.

[2001] Einstein meets Hilbert: At the crossroads of physics and mathematics, *Physics in Perspective*, 3 (2001), pp. 369–424.

[2002] Einstein's gravitational field equations and the Bianchi identities, *Mathematical Intelligencer*, 24(4) (2002), pp. 57–66.

[2009] A look back at Hermann Minkowski's Cologne lecture "Raum und Zeit," *The Mathematical Intelligencer*, 31(2) (2009), pp. 27–39.

RUMER (Georg B.)

[1929] Über eine Erweiterung des allgemeinen Relativitätstheorie, *Göttinger Nachrichten*, 1929, pp. 92–99.

[1931a] Zur allgemeinen Relativitätstheorie, *Göttinger Nachrichten*, 1931, pp. 148–156.

[1931b] Der gegenwärtige Stand der Diracschen Theorie des Elektrons, *Physikalische Zeitschrift*, 32 (1931), pp. 601–622.

RUND (Hanno)

[1966] *The Hamilton–Jacobi Theory in the Calculus of Variations: Its Role in Mathematics and Physics*, London, New York: D. Van Nostrand, 1966.

RZEWUSKI (Jan)

[1953] Differential conservation laws in non-local field theories, *Il Nuovo Cimento* IX, 10 (1953), pp. 784–802.

[1958] *Field Theory, I: Classical Theory*, Warsaw: PWN, Polish Scientific Publishers, 1958; reprints, London: Iliffe Books, 1964, 1967.

SAGAN (Hans)

[1969] *Introduction to the Calculus of Variations*, New York: McGraw-Hill, [1969].

SARDANASHVILY (Gennadi)

[2005] Noether identities of a generic differential operator. The Koszul-Tate complex, *International Journal of Geometric Methods in Modern Physics*, 2 (2005), pp. 873–886.

SAUER (Tilman)

[1999] The relativity of discovery: Hilbert's first note on the foundation of physics, *Archive for History of Exact Sciences*, 53 (1999), pp. 529–575.

[2006] Field equations in teleparallel spacetime: Einstein's *Fernparallelismus* approach toward unified field theory, *Historia Matematica*, 33(4) (2006), pp. 399–439.

SCHIFF (Leonard Isaac)

[1949] *Quantum Mechanics*, New York, Toronto, London: McGraw-Hill Book Co., 1949.

SCHLOTE (Karl-Heinz)

[1991] Fritz Noether—Opfer zweier Diktaturen, Tod und Rehabilitierung, *NTM, Schriftenreihe für Geschichte der Naturwissenschaften, Technik, und Medizin*, 28 (1991), pp. 33–41.

SCHOLZ (Erhard)

[1999a] The concept of manifold, 1850–1950, in *History of Topology*, Ioan Mackenzie James, ed., Amsterdam: North-Holland, 1999, pp. 25–64.

[1999b] Weyl and the theory of connections, in Gray [1999], pp. 260–284.

SCHOLZ (Erhard), ed.

[2001] *Hermann Weyl's Raum-Zeit-Materie and a General Introduction to His Scientific Work*, DMV Seminar 30, Basel: Birkhäuser, 2001.

SCHOUTEN (Jan Arnoldus)
[1924] *Der Ricci-Kalkül*, Berlin: Springer, 1924; reprint, 1978.
[1951] *Tensor Analysis for Physicists*, Oxford: Clarendon Press, 1951; 2nd ed., 1954; correctied printing, 1959; reprint, New York: Dover, 1989.
[1954] *Ricci Calculus*, 2nd ed., Berlin, Göttingen, Heidelberg: Springer, 1954.

SCHÜTZ (Ignaz R.)
[1897] Prinzip der absoluten Erhaltung der Energie, *Göttinger Nachrichten* (1897), Part 2, pp. 110–123.

SEGAL (Sanford L.)
[2003] *Mathematicians under the Nazis*, Princeton and Oxford: Princeton University Press, 2003.

ŚLEBODZIŃSKI (Władisław)
[1931] Sur les équations canoniques de Hamilton, *Académie royale de Belgique, Bulletins de la classe des sciences*, 5th series, 17 (1931), pp. 864–870.

SMALE (Stephen)
[1970] Topology and mechanics I, *Inventiones Mathematicae*, 10 (1970), pp. 305–331.

SMITH (Larry)
[2000] Noether's bound in the invariant theory of finite groups and vector invariants of iterated wreath products of symmetric groups, *Quarterly Journal of Mathematics*, 51 (2000), pp. 93–105.

ŚNIATYCKI (Jedrzej)
[1970] On the geometric structure of classical field theory in Lagrangian formulation, *Mathematical Proceedings of the Cambridge Philosophical Society*, 68 (1970), pp. 475–484.

SOURIAU (Jean-Marie)
[1957] Le tenseur impulsion-énergie en relativité variationnelle, *Comptes rendus de l'Académie des sciences de Paris*, 245 (1957), pp. 958–960.
[1964] *Géométrie et relativité*, Paris: Hermann, 1964.
[1966] Quantification géométrique, *Communications in Mathematical Physics*, 1 (1966), pp. 374–398.
[1970] *Structure des systèmes dynamiques*, Paris: Dunod, 1970; reprint, Paris Éditions Jacques Gabay, 2008; English translation by C. H. Cushman-de Vries, *Structure of Dynamical Systems. A Symplectic View of Physics*, Springer, 1997.
[1974] Sur la variété de Képler, in *Symposia Mathematica*, vol. 14, London, New York: Academic Press, 1974, pp. 343–360.

SRINIVASAN (Bhama), SALLY (Judith D.), eds.
[1983] *Emmy Noether in Bryn Mawr*, New York, Berlin, Tokyo: Springer-Verlag, 1983.

STACHEL (John)
[1989] Einstein's search for general covariance, 1912–1925, in Howard and Stachel [1989], pp. 62–100.
[1994] Lanczos's early contributions to relativity and his relationship with Einstein, in *Proceedings of the Cornelius Lanczos International Centenary Conference* (Raleigh, NC, 1993), J. David Brown, Moody T. Chu, Donald C. Ellison and Robert J. Plemmons, eds., Philadelphia, PA: Society for Industrial and Applied Mathematics, 1994, pp. 201–221.

STASHEFF (James D.)
[1997] Deformation theory and the Batalin-Vilkovisky master equation, in *Deformation Theory and Symplectic Geometry* (Ascona, 1996), Daniel Sternheimer, John Rawnsley, and Simone Gutt, eds., Dordrecht: Kluwer, 1997.
[2005] Poisson homotopy algebra: An idiosyncratic survey of homotopy algebraic topics related to Alan's interests, in *The Breadth of Symplectic and Poisson Geometry, Festschrift in Honor of Alan Weinstein*, Jerrold E. Marsden and Tudor S. Ratiu, eds., Progress in Mathematics, vol. 232, Boston, MA: Birkhäuser, 2005, pp. 583–601.

STEUDEL (Heinz)

[1965] Invariance groups and conservation laws in linear field theories, *Nuovo Cimento*, 39 (1965), pp. 395–397.

[1966] Die Struktur der Invarianzgruppe für lineare Feldtheorien, *Zeitschrift für Naturforschung*, 21 A (1966), pp. 1826–1828.

[1967] Zur Umkehrung der Noetherschen Satzes, *Annalen der Physik*, 7th series, 20 (1967), pp. 110–112.

STUBHAUG (Arild)

[2000] *Det var mine tankers djervhet—Matematikeren Sophus Lie*, Oslo: H. Aschehoug and Co., 2000; English translation by Richard H. Daly, *The Mathematician Sophus Lie: It was the audacity of my thinking*, Berlin, Heidelberg: Springer, 2002; French translation by Marie-José Beaud and Patricia Chwat, *Sophus Lie. Une pensée audacieuse*, Paris, Berlin, Heidelberg, New York: Springer, 2006.

STUDY (Eduard)

[1923] *Einleitung in die Theorie der Invarianten linearer Transformationen auf Grund der Vektorenrechnung*, Braunschweig: Vieweg, 1923.

SYLVESTER (James Joseph)

The Mathematical Collected Papers, vol. 1, Cambridge: Cambridge University Press, 1904; corrected reprint, New York: Chelsea, 1973.

[1851] On the general theory of associated algebraical forms, *Cambridge and Dublin Mathematical Journal*, 6 (1851), pp. 289–293; *The Mathematical Collected Papers*, vol. 1, pp. 198–202.

TAKENS (Floris)

[1979] A global formulation of the inverse problem of the calculus of variations, *Journal of Differential Geometry*, 14 (1979), pp. 543–562.

TATON (René)

[1983] Les relations d'Euler avec Lagrange, in *Leonhard Euler: 1707–1783. Beiträge zu Leben und Werk* Basel, Boston: Birkhäuser Verlag, 1983, pp. 409–420; reprinted in René Taton, *Études d'histoire des sciences*, recueillies pour son 85ᵉ anniversaire par Danielle Fauque, Myriana Ilic et Robert Halleux, Turnhout: Brepols, 2000, pp. 273–284.

TAVEL (Morton A.)

[1971a] Milestones in mathematical physics: Noether's theorem, *Transport Theory and Statistical Physics*, 1(3) (1971), pp. 183–185.

[1971b] Application of Noether's theorem to the transport equation, *Transport Theory and Statistical Physics*, 1(4) (1971), pp. 271–285.

TEICHER (Mina), ed.

[1999] *The Heritage of Emmy Noether*, Israel Mathematical Conference Proceedings, vol. 12, [Ramat-Gan]: Bar-Ilan University, 1999.

THIELE (Rüdiger)

[1997] On some contributions to field theory in the calculus of variations from Beltrami to Carathéodory, *Historia Mathematica*, 24 (1997), pp. 281–300.

THOMSON (Joseph John)

[1888] *Applications of Dynamics to Physics and Chemistry*, London: Macmillan, 1888.

T'HOOFT (Gerardus), ed.

[2005] *50 Years of Yang–Mills Theory*, New Jersey, London, Singapore: World Scientific, 2005.

TOBIES (Renate)

[2001] Von Sofja Kowalewskaja bis Emmy Noether. Frauenkarrieren in der Mathematik im Vergleich mit Männerkarrieren, *Sitzungsberichte der Berliner mathematischen Gesellschaft* (2001), pp. 537–545.

TOLLMIEN (Cordula)
[1991] Die Habilitation von Emmy Noether an der Universität Göttingen, *NTM, Schriftenreihe für Geschichte der Naturwissenschaften, Technik, und Medizin*, 28 (1991), pp. 13–32.

TONELLI (Leonida)
[1921] *Fondamenti di calcolo delle variazioni*, 2 vol., Bologna: Nicola Zanichelli, 1921.

TONTI (Enzo)
[1969] Variational formulation of nonlinear differential equations, (I) and (II), *Académie royale de Belgique, Bulletins de la classe des sciences*, 5th series, 55 (1969), pp. 137–165 and 262–278.

TRAUTMAN (Andrzej)
[1956] On the conservation theorems and equations of motion in covariant field theories, *Bulletin de l'Académie polonaise des sciences, Série des sciences mathématiques, astronomiques et physiques* Cl. III, 4 (1956), pp. 675–678.
[1957] On the conservation theorems and coordinate systems in general relativity, *Bulletin de l'Académie polonaise des sciences, Série des sciences mathématiques, astronomiques et physiques* Cl. III, 5 (1957), pp. 721–727.
[1962] Conservation laws in general relativity, in *Gravitation, An Introduction to Current Research*, Louis Witten, ed., New York: Wiley, 1962, pp. 169–198.
[1967] Noether equations and conservation laws, *Communications in Mathematical Physics*, 6 (1967), pp. 248–261.
[1972] Invariance of Lagrangian Systems, in *General Relativity*, papers in honor of J. L. Synge, L. O'Raifeartaigh, ed., Oxford: Clarendon Press, 1972, pp. 85–99.

TREIMAN (Sam B.), JACKIW (Roman W.), and GROSS (David J.)
[1972] *Lectures on Current Algebra and Its Applications*, Princeton: Princeton University Press, 1972.

TRESSE (Arthur)
[1893] Sur les invariants différentiels des groupes continus de transformations, *Acta Mathematica*, 18 (1893), pp. 1–88.

TSUJISHITA (Toru)
[1982] On variation bicomplexes associated to differential equations, *Osaka Journal of Mathematics*, 19 (1982), pp. 311–363.

TULCZYJEW (Włodzimierz M.)
[1975] Sur la différentielle de Lagrange, *Comptes rendus de l'Académie des sciences de Paris*, 280 (1975), pp. 1295–1298.
[1977] The Lagrange complex, *Bulletin de la Société mathématique de France*, 105 (1977), pp. 419–431.

UHLENBECK (Karen)
[1983] Conservation laws and their application in global differential geometry, in Srinivasan and Sally [1983], pp. 103–115.

VAN DANTZIG (David)
[1932] Zur allgemeinen projektiven Differentialgeometrie, I and II, *Koninklijke Akademie van Wetenschappen te Amsterdam, Proceedings of the Section of Sciences*, 35 (1932), pp. 524–534 and 535–542.

VAN DER WAERDEN (Bartel L.)
[1932] *Die gruppentheoretische Methode in der Quantenmechanik*, Berlin: J. Springer, 1932; English translation, *Group Theory and Quantum Mechanics*, Berlin: Springer, 1974.
[1935] Nachruf auf Emmy Noether, *Mathematische Annalen*, 111 (1935), pp. 469–476.

VAN NIEUWENHUIZEN (Peter)
[1981] Supergravity, *Physics Reports*, 68 (4) (1981), pp. 189–398.

VERMEIL (Hermann)

[1919] Bestimmung einer quadratischen Differentialform aus den Riemannschen und den Christoffelschen Differentialinvarianten mit Hilfe von Normalkoordinaten, *Mathematische Annalen*, 79 (1919), pp. 289–312.

VINOGRADOV (Alexander M.)

[1977] On the algebro-geometric foundations of Lagrangian field theory [in Russian], *Doklady Akad. Nauk SSSR*, 236 (1977), pp. 284–287; English translation, *Soviet Mathematics Doklady*, 18 (1977), pp. 1200–1204.

[1979] The theory of higher infinitesimal symmetries of nonlinear partial differential equations [in Russian], *Doklady Akad. Nauk SSSR*, 248 (1979), pp. 274–278; English translation, *Soviet Mathematics Doklady*, 20 (1979), pp. 985–990.

[1984a] Local symmetries and conservation laws, *Acta Applicandae Mathematicae*, 2 (1984), pp. 21–78.

[1984b] The $\mathscr{C}$-spectral sequence, Lagrangian formalism, and conservation laws, I and II, *Journal of Mathematical Analysis and Applications*, 100 (1984), pp. 1–40 and 41–129.

VIZGIN (Vladimir P.)

[1972] *Razvitije vzaimosvyazi principov invariantnosti s zakonami sokhranenija v klassicheskoj fizike* [Russian] (Development of the Interconnection between Invariance Principles and Conservation Laws in Classical Physics), Moscow: Nauka, 1972.

[1985] *Edinye teorii polja v perevoj treti XX veka* [Russian], Moscow: Nauka, 1985; English translation by Julian B. Barbour, *Unified Field Theories in the First Third of the 20th Century*, Science Networks, Historical Studies, vol. 13, Basel: Birkhäuser, 1994.

VOLTERRA (Vito Isacar)

Opere matematiche, Volume primo 1881–1892, Roma: Accademia Nazionale dei Lincei, 1954.

[1887] Sopra le funzioni che dipendono da altre funzioni, Nota I, *Atti della Reale Accademia dei Lincei. Rendiconti*, Ser. IV, vol. III (1887), pp. 97–105; *Opere matematiche*, vol. 1, pp. 294–302.

[1913] *Leçons sur les fonctions de ligne* : professées à la Sorbonne en 1912, Joseph Pérès, rédacteur, Paris: Gauthier-Villars, 1913.

WEIL (André)

[1991] *Souvenirs d'apprentissage*, Basel, Boston: Birkhäuser, 1991; English translation by Jennifer Gage, *The Apprenticeship of a Mathematician*, Basel, Boston, Berlin: Birkhäuser, 1992.

WEINSTEIN (Alan D.)

[2005] The geometry of momentum, in *Géométrie au XXe siècle. Histoire et horizons*, Joseph Kouneiher, Dominique Flament, Philppe Nabonnand and Jean-Jacques Szczeciniarz, eds., Paris: Hermann, 2005, pp. 236–245.

WEITZENBÖCK (Roland)

[1922] Differentialinvarianten, in *Encyclopädie der mathematischen Wissenschaften* III.3 (Geometrie), Leipzig: B. G. Teubner, 1922, pp. 1–71.

[1923] *Invariantentheorie*, Groningen: Noordhoff, 1923.

WENTZEL (Gregor)

[1943] *Einführung in die Quantentheorie der Wellenfelder*, Vienna: Franz Deuticke, 1943; English translation by Charlotte Houtermans and Josef M. Jauch, *Quantum Theory of Fields*, New York: Interscience, 1949; reprint, New York: Dover, 2003.

[1957] *Lectures on Special Topics in Field Theory*, Bombay: Tata Institute of Fundamental Research, 1957.

WEYL (Hermann)

Selecta, Basel: Birkhäuser, 1956.

Gesammelte Abhandlungen, 4 vol., K. Chandrasekharan, ed., Berlin: Springer-Verlag, 1968.

[1917] Zur Gravitationstheorie, *Annalen der Physik* (Leipzig) (4), 54 (1917), pp. 117–145; *Gesammelte Abhandlungen*, vol. 1, pp. 670–698.

[1918a] Gravitation und Elektrizität, *Sitzungsberichte der Königlich-preussischen Akademie der Wissenschaften zu Berlin*, 26 (1918), pp. 465–480 (meeting of 2 May 1918); reprinted in Lorentz *et al.* [1922], 5th ed., 1923, pp. 147–159; *Selecta*, pp. 179–192; *Gesammelte Abhandlungen*, vol. 2, pp. 29–42 (with a commnentary by Einstein and an additional text by the author written in 1955); English translations, Gravitation and electricity, in Lorentz *et al.* [1923], pp. 199–216, and in O'Raifeartaigh [1997], pp. 24–37.

[1918b] *Raum, Zeit, Materie*, Berlin: J. Springer, 1918; 2nd ed., 1919; 3rd ed., revised, 1919; 4th ed., revised, 1921; 5th ed., unchanged, 1923; 6th ed., 1977; English translation by H. L. Brose based on the 4th German ed., *Space, Time, Matter*, London: Methuen, 1922; reprint, New York: Dover, 1952; French translation by G. Juvet and R. Leroy, based on the 4th German ed., *Temps, Espace, Matière, Leçons sur la théorie de la relativité générale*, Paris: A. Blanchard, 1922; reprint, 1958, with commentaries by Georges Bouligand.

[1918c] Reine Infinitesimalgeometrie, *Mathematische Zeitschrift*, 2 (1918), pp. 384–411, *Gesammelte Abhandlungen*, vol. 2, pp. 1–28.

[1919] Eine neue Erweiterung der Relativitätstheorie, *Annalen der Physik*, 59 (1919), pp. 101–133; *Gesammelte Abhandlungen*, vol. 2, pp. 55–87.

[1928] *Gruppentheorie und Quantenmechanik*, Leipzig: S. Hirzel, 1928; 2nd ed., revised, 1931; English translation by H. P. Robertson, *The Theory of Groups and Quantum Mechanics*, New York: Dutton, 1932; reprint, New York: Dover, 1949.

[1929] Elektron und Gravitation, *Zeitschrift für Physik*, 56 (1929), pp. 330–352; *Gesammelte Abhandlungen*, vol. 3, pp. 245–267; English translation, Electron and gravitation, in O'Raifeartaigh [1997], pp. 121–144.

[1935a] Emmy Noether (Memorial address delivered in Goodhart Hall, Bryn Mawr College, on April 26, 1935), *Scripta Mathematica*, 3 (1935), pp. 201–220; *Gesammelte Abhandlungen*, vol. 3, pp. 425–444.

[1935b] Geodesic fields in the calculus of variation for multiple integrals, *Annals of Mathematics*, 36 (1935), pp. 607–629; *Gesammelte Abhandlungen*, vol. 3, pp. 470–492.

[1939] *The Classical Groups*, Princeton: Princeton University Press, 1939; reprint, 1946; 2nd ed., revised, 1953; reprint, 1997.

[1944] David Hilbert and his mathematical work, *Bulletin of the American Mathematical Society*, 50 (1944), pp. 612–654; *Gesammelte Abhandlungen*, vol. 4, pp. 130–172.

WHITTAKER (Edmund Taylor)

[1904] *A Treatise on the Analytical Dynamics of Particles and Rigid Bodies*, Cambridge: at the University Press, 1904; 2nd ed., 1917; German translation by F. and K. Mittelsten Scheid, Berlin: Springer, 1924; 3rd ed., Cambridge, 1927; 4th ed., Cambridge, 1937; 5th ed. and first American ed., New York: Dover, 1944.

WIGNER (Eugene)

The Collected Works of Eugene Paul Wigner, Part A. The Scientific Papers, vol. 1, Arthur S. Wightman, Jagdish Mehra, Brian R. Judd and George W. Mackey, eds., Berlin, Heidelberg: Springer-Verlag, 1993.

The Collected Works of Eugene Paul Wigner, Part B. Historical, philosophical, and sociopolitical papers, vol. 6, *Philosophical Reflections and Syntheses*, with notes by Gérard G. Emch; Jagdish Mehra and Arthur S. Wightman, eds., Berlin: Springer-Verlag, 1997; reprint of *Philosophical Reflections and Syntheses*, Berlin: Springer-Verlag, 1995.

[1927] Über die Erhaltungssätze in der Quantenmechanik, *Göttinger Nachrichten* (1927), pp. 375–381; *The Collected Works of Eugene Paul Wigner*, Part A, vol. 1, pp. 84–90.

[1931] *Gruppentheorie und ihre Anwendung auf die Quantenmechanik der Atomspektren*, Braunschweig: F. Vieweg & Sohn, 1931; English translation, *Group Theory and Its Applications to the Quantum Mechanics of Atomic Spectra*, New York: Academic Press, 1959.

[1949] Invariance in physical theory, *Proceedings of the American Philosophical Society*, 93 (7) (December 1949), pp. 521–526; reprinted in [1967], pp. 3–13; *The Collected Works of Eugene Paul Wigner*, Part B, vol. 6, pp. 283–293.

[1954] Conservation laws in classical and quantum physics, *Progress of Theoretical Physics*, 11 (1954), pp. 437–440.

[1964a] Symmetry and conservation laws, *Proceedings of the National Academy of Sciences*, 51 (5) (May 1964), pp. 956–965; reprinted in *Physics Today*, 17 (1964), pp. 34–40 and in [1967], pp. 14–27; *The Collected Works of Eugene Paul Wigner*, Part B, vol. 6, pp. 295–310.

[1964b] Events, laws of nature, and invariance principles, in *The Nobel Prize Lectures*, The Nobel Foundation, Amsterdam, London, New York: Elsevier Publishing Company, 1964; reprinted in *Nobel Lectures, Physics, 1963–1970*, Amsterdam, London, New York: Elsevier Publishing Company, 1972, pp. 6–17 and in [1967], pp. 38–50; *The Collected Works of Eugene Paul Wigner*, Part B, vol. 6, pp. 321–333.

[1967] *Symmetries and Reflections, Scientific Essays*, Bloomington, London: Indiana University Press, 1967.

[1995] *Events, laws of nature, and invariance principles*, mimeographed notes, ca. 1980, published in *The Collected Works of Eugene Paul Wigner*, Part B, vol. 6, pp. 334–342.

WINOGRADZKI (Judith)
[1956] Sur le tenseur impulsion-énergie métrique et le théorème de Noether, *Cahiers de Physique*, 10th series, no. 67 (1956), pp. 1–5.

WINTNER (Aurel)
[1941] *The Analytical Foundations of Celestial Mechanics*, Princeton Mathematical Series, vol. 5, Princeton: Princeton University Press, 1941; reprint, 1947.

WRIGHT (Joseph Edmund)
[1908] *Invariants of Quadratic Differential Forms*, Cambridge Tracts in Mathematics and Mathematical Physics, vol. 9, Cambridge: at the University Press, 1908; reprint, New York: Hafner Publishing Company, *s.d.*

YANG (Chen Ning)
Selected Papers, 1945–1980, With Commentary, San Francisco: W. H. Freeman and Co., 1983.
[1957] The law of parity conservation and other symmetry laws of physics, in *Les Prix Nobel*, Stockholm: The Nobel Foundation, 1957, pp. 95–105; reprinted in *Science*, 127 (1958), pp. 565–569; reprinted in *Nobel Lectures, Physics 1942–1962*, Amsterdam, London, New York: Elsevier Publishing Company, 1964, pp. 393–403; *Selected Papers*, pp. 236–246.

[1980a] Einstein's impact on theoretical physics, *Physics Today*, 33 (6) (June 1980), pp. 42–44 and 48–49; *Selected Papers*, pp. 563–567.

[1980b] Fibre bundles and the physics of the magnetic monopole, in *The Chern Symposium 1979* (Proceedings of the international symposium on differential geometry in honor of S. S. Chern, Berkeley, CA, 1979), Wu-Yi Hsiang, Shoshichi Kobayashi, Isadore M. Singer, Alan Weinstein, Joseph Wolf and Hung Hsi Wu, eds., New York, Berlin: Springer-Verlag, 1980, pp. 247–253.

[1986] Hermann Weyl's contribution to physics, in Chandrasekharan [1986], pp. 7–21.

ZUCKERMAN (Gregg J.)
[1987] Action principles and global geometry, in *Mathematical Aspects of String Theory* (University of California, San Diego, 1986), S. T. Yau, ed., Advanced Series in Mathematical Physics, vol. 1, Singapore, New Jersey, Hong Kong: World Scientific, 1987, pp. 259–284.

Index

Printed in the United States
By Bookmasters